Max Weber and Charles Peirce

Max Weber and Charles Peirce

At the Crossroads of Science, Philosophy, and Culture

Basit Bilal Koshul

LEXINGTON BOOKS
Lanham • Boulder • New York • Toronto • Plymouth, UK

Published by Lexington Books
A wholly owned subsidiary of Rowman & Littlefield
4501 Forbes Boulevard, Suite 200, Lanham, Maryland 20706
www.rowman.com

10 Thornbury Road, Plymouth PL6 7PP, United Kingdom

British Library Cataloguing in Publication Information Available

Library of Congress Cataloging-in-Publication Data

Koshul, Basit Bilal, 1968-
Max Weber and Charles Peirce : at the crossroads of science, philosophy and culture / Basit Koshul.
pages cm
Includes bibliographical references and index.
ISBN 978-0-7391-7800-3 (cloth : alk. paper)— ISBN 978-0-7391-7801-0 (electronic)
1. Weber, Max, 1864-1920. 2. Peirce, Charles S. (Charles Sanders), 1839-1914. 3. Science—Philosophy. 4. Social sciences—Philosophy. 5. Culture—Philosophy. I. Title.
HM479.W42K667 2014
301.092'2—dc23

2013025866

™ The paper used in this publication meets the minimum requirements of American National Standard for Information Sciences Permanence of Paper for Printed Library Materials, ANSI/NISO Z39.48-1992.

Printed in the United States of America

In memory of Otto Maduro,
April 14, 1945–May 9, 2013

Contents

Abbreviations of Weber's Works

CIS 1981. Some categories of interpretive sociology. *Sociological Quarterly*, *22*, 151–180.

CS 1977. *Critique of Stammler*. (Guy Oakes, Trans.). New York, NY: Free Press.

ES 1968. *Economy and society*. Guenther Roth & Claus Wittich (Eds.). New York, NY: Bedminster.

LCS 1949. Critical studies in the logic of the cultural sciences: A critique of Eduard Meyer's methodological views. In Edward A. Shils & Henry A. Finch (Eds.), *Max Weber on the methodology of the social sciences* (pp. 113–188). Glencoe, IL: Free Press.

MEN 1949. The meaning of 'ethical neutrality.' In Edward A. Shils & Henry A. Finch (Eds.), *Max Weber on the methodology of the social sciences* (pp. 1–47). Glencoe, IL: Free Press.

OSS 1949. 'Objectivity' in social science and social policy. In Edward A. Shils & Henry A. Finch (Eds.), *Max Weber on the methodology of the social sciences* (pp. 49–112). Glencoe, IL: Free Press.

PESC 2002. *The protestant ethic and the spirit of capitalism.* (Stephen Kalberg, Trans.). Los Angeles, CA: Roxbury Press.

RK 1975. *Roscher and Knies: The logical problems of historical economics.* (Guy Oakes, Trans.). New York, NY: Free Press.

RRW 1946. Religious rejections of the world and their directions. In H. H. Gerth & C. Wright Mills (Eds.), *From Max Weber* (pp. 323–359). New York, NY: Oxford University Press.

SPWR 1946. The social psychology of world religions. In H. H. Gerth & C. Wright Mills (Eds.), *From Max Weber* (pp. 267–322). New York, NY: Oxford University Press.

SV 1946. Science as a vocation. In H. H. Gerth & C. Wright Mills (Eds.), *From Max Weber* (pp. 129–156). New York, NY: Oxford University Press.

Acknowledgments

Firstly:

> *All praise and glory are due to Allah who has guided us to this destination, for we would not have been able to guide ourselves if Allah had not guided us.*
>
> —Qur'an, 7:34

Thereafter:

The completion of this project is the result of support, patience, and encouragement from two other quarters—my family and my teachers. All that I have accomplished thus far and anything that I accomplish in the future is the result of the prayers and sacrifices of my parents, Muhammad Ikram Koshul and Shagufta Ikram Koshul. The best I can do to "repay" them is to offer the following prayer: "My Lord! Shower Your grace upon them both, just as they cherished and reared me while I was a child!" (Qur'an, 17:25). After the support of my parents, the patience and dedication of my wife, Samia Nazneen Tabassum, has been the most important pillar of support. She has been more than just a helper—she has been a teacher. The lessons I have learned from her are an essential part of the "spirit" of this text. Without her contribution the "spirit" of the text would have been very different. I extend gratitude also to my children—Falucq, Firyal, Fariha, Ibrahim, and Ismail—for putting up with the difficulties and depravations that come with a father who spent most of their childhood enrolled in school.

With the same amount of joy and gratitude that comes with acknowledging the support of my family, I acknowledge the debt I owe to my teachers. With respect to this particular project, I am most indebted to Peter Ochs. Of all the teachers with whom I have had a relationship (either formal or informal), the relationship with Prof. Ochs has been the most significant—both quantitatively and qualitatively. Beginning with introducing me to the work of Charles Peirce around 1997 at Drew University, through the writing of my dissertation at the University of Virginia, and during the further development and revision of the thesis, Prof. Ochs has been a demanding mentor, a perceptive guide, and a gracious friend. If it were not for this relationship and the pioneering work that he did in the area of Scriptural Reasoning, I would have left academia a long time ago. It is with a heavy heart that I acknowledge the debt that I owe to Otto Maduro who passed away as I was in the final stages of completing this manuscript. Prof. Maduro was as critical, demanding, and supportive in my study of the sociology of religion and Max Weber as Prof. Ochs has been in my study of the philosophy of religion and Peirce. In dedicating this book to his memory I am making a feeble and inadequate attempt to express the depth of my gratitude to him. I am also indebted to Chuck Mathews and Jamie Ferreira—outstanding scholars and outstanding human beings. The time I spent in their classrooms and the conversations that I had with them outside the classroom are enduring memories from my stay in Virginia.

A special note of gratitude to Mohammed Azam Khan and Shahida Azam Khan and their four wonderful children, Nawal, Daniyal, Misha, and Malaika. I spent three months as a guest at their home in Ruckersville, Virginia, during the final stage of writing my dissertation. For the duration of the three months, they were gracious hosts, always attending to my wants before taking care of their own needs. The Ramadan of 2010 that I spent with them will remain a cherished memory long into the future.

During the writing and revision of the project, I was fortunate to have the help of Ahmed Afzaal in bibliographical searches and in proofreading different chapters. Thanks to Nida Haseeb Khan and Zainuddin Moulvi for proofreading. Special thanks to Mian Muhammad Nauman Faizi for not only helping with the "grunt work" but also for intense conversations that helped to clarify and crystallize important parts of the argument that is made in chapter five.

I would be remiss to not acknowledge the gratitude that is owed to Jana Hodge-Kluck, the associate editor for Philosophy, Classics, Sociology, and Criminology at Lexington Books. Her professionalism during the process of submission, review, and revision of the manuscript was exceeded only by her patience and understanding.

Introduction

Max Weber saw the rule of law, liberal democracy, free trade, scientific inquiry, and open horizons as the defining ideals of modern Western culture. He devoted the best years of his life to studying the origins, developments, and trajectory of these ideals during the course of history. A succinct summary of his research question would be: "In their underdeveloped form, these ideals are found in cultures across the globe. In their most developed form they are found in only modern, Western. What are the factors that caused a universally latent potential to become actualized at a particular place and particular time in history?" He did not pursue this line of inquiry for the sake of demonstrating the superiority of Western culture above others, nor because he had any doubts about the value of these ideals, nor because he needed to find a topic that would come under the heading of "original contribution to scholarship." And he certainly did not undertake this study because he needed to rack up the numbers of his publications so that he could earn tenure. Weber pursued this line of inquiry because he was deeply committed to these ideals and was acutely worried about their vitality in the coming decades. Among his contemporaries and predecessors, there were a variety of theories that located the origins of these ideals in the "laws of history," "laws of nature," something inherent in "human nature," or something inherent in the nature of Western culture. He rejected all of these theories. For Weber these ideals were the product of a unique constellation of historical factors at a particular place and a particular time in history. This constellation opened up a range of future cultural possibilities. Among the most important factors to determine which of the possibilities were actualized in history were certain choices and decisions made by cultural beings who were a part of this constellations.

By the end of the nineteenth century, it was clear that the constellation of factors in the midst of which the defining ideals of Western culture had originally emerged had fallen apart and was being replaced by a new constellation. The emergence of this new constellation was radically changing the range of future cultural possibilities along with the character of the cultural beings living in the midst of this new constellation. The new range of cultural possibilities and the emergent character of the cultural beings that the new constellation of historical factors was producing did not bode well for the health and vitality of the rule of law, liberal democracy, free trade, scientific inquiry, and open horizons in the twentieth century and beyond.

The crisis that Weber saw unfolding in his time has continued to progress and manifests itself in somewhat different forms at the beginning of the twenty-first century. From the perspective of certain developments in post-Weber scholarship, the contemporary manifestation of the cultural crisis that caused Weber so much angst is expressed in three different but interrelated crises. The three interrelated crises are the crisis in modern science (or the question "what is science?"), the crisis in the modern social sciences (or the question "are the soft sciences scientific?"), and the crisis in modern culture (or the question "can we affirm meaning (*Sinn*) in the face of scientific knowledge?"). This book is about the relevance of two thinkers from the not-so-distant past in helping us to better understand these crises so that we are better equipped to redress them. In addition to the insights of Weber (d. 1920), the work of Charles Peirce (d. 1914) will help us to navigate the issues raised above. The insights of three post-Weber scholars—Walker Percy, Peter Berger, and Robert Bellah—help us to see the three crises in some detail and allow us to offer the preliminary hypothesis that Weber and Peirce still speak to us today.

Walker Percy argues that it is impossible to escape the conclusion that modern science is in the midst of a profound and paradigmatic crisis. The evidence for this conclusion comes from the paradoxical and contradictory results of the increase of scientific knowledge over the past three hundred years. The following picture emerges if we divide science into two categories—the hard sciences and the soft sciences. On the one hand the area of ignorance about physical-natural reality has receded steadily as the hard sciences have advanced. On the other hand the development of the soft sciences has been accompanied by increasing incoherence in knowledge about that which is uniquely human. Percy notes: "Modern science is itself radically incoherent, not when it seeks to understand things and subhuman organisms and the cosmos itself, but when it seeks to understand man, not man's physiology or neurology or his bloodstream, but man *qua* man, man when he is peculiarly human" (Percy, 1991, 271). The one characteristic that makes the human being unique among all natural and physical phenomena is the fact that the human being is a heterogeneous composite of what are often called "mind" and "body." When Percy says that modern science is "radically incoherent" he means something very specific—modern science has no coherent account of what it means to be human because it has not been able to close the gap between mind and body opened up by Descartes at the dawn of modern philosophy. Referring to the gap between mind and body, Percy notes: "I refer to this gap in scientific knowledge as an incoherence, from the Latin *in-cohaerere*, a not-sticking together" (Percy, 1991, 275). Percy argues that the crisis in modern science manifests itself most clearly in the soft sciences: "In the case of the cosmos there is the sense that the areas of ignorance are being steadily eroded by the advance of science. In the case of the sciences of man, however, the incoherence is chronic and seems to be intractable" (Percy,1991, 273). For Percy this incoherence in scientific knowledge has put the very integrity of the scientific enterprise at stake.

Closely related to the crisis of incoherence in modern science is the crisis in the soft sciences. The soft sciences have sought to replicate the success of the hard sciences by applying the tools and concepts of the hard sciences to the study of being human and social reality. After more than a century of this mimicry, the results are plainly visible. Speaking from within a particular discipline, sociology, Peter Berger argues that "parochialism, triviality, rationalism, and ideology" have become the dominant characteristics of contemporary sociology and there are clear signs that the discipline is in very poor health (Berger, 1992, 16). Berger notes that two of the reasons for the prevailing condition of sociology (triviality and rationalism) are directly related to continued attempts to mimic the hard sciences in the study of the human. For much of the post-WWII period, we find that in "a futile and theoretically misguided attempt to ape the natural sciences sociologists developed ever more refined quantitative methods of research" (Berger, 1992, 17). Second, just as the natural scientists built mathematical models to map natural phenomena like the weather, tectonic shifts, migration of birds and animals, sociologists, political scientists, economists, etc., have sought to map certain uniquely human characteristics (i.e., rationality, religious belief, value commitments) in roughly equivalent quantifiable terms. Berger argues that the "pathology now goes very deep indeed" so much so that sociology is on the verge of becoming "obsolete" (Berger, 1992, 18) in the near future—if it has not already become so. In offering this prognosis for his own field, Berger notes that "one should not view it in isolation. Its symptoms tend to be those afflicting the intellectual life in general. Other human sciences are in no better shape" (Berger, 1992, 18). When we look at Berger's analysis of sociology and the social sciences in light of Percy's observations, we can draw the following inference. The single most important factor contributing to the crisis in the soft sciences is the use of the tools and concepts of the hard sciences to study and understand the uniquely human.

The crisis in the soft sciences has implication far outside the walls of the university and the halls of academia. Robert Bellah argues that there is a direct link between the crisis in the soft sciences and a deep disquiet in modern culture. Speaking as a sociologist and in line with the insights of Weber and Durkheim, Bellah offers the following evaluation of the health of contemporary culture: "The adequacy of any ultimate perspective is its ability to transform human experience so that it yields life instead of death. Our present fragmented and disorganized culture does not rank high on that criterion" (Bellah, 1991, 245). Bellah argues that the root cause of the "fragmented and disorganized" condition of modern culture is that (with greater fervor and zeal than the hard sciences) the soft sciences have advocated an Enlightenment myth "which views science as the bringer of light relative to which religion and other dark things will vanish away" (Bellah, 1991, 238). This modern incarnation of Manichean thought sees intuitive, emotional, imaginative, and subjective elements (epitomized in religion) as the dark forces against which war has to be waged by the forces of light represented by cognition, reason, rationality, and objectivity (epitomized in science). The attraction of this Manichean myth has been so powerful that it has seduced even notable modern theologians who have formulated "a cognitive conception of religious belief that makes it parallel to objectivist scientific description" (Bellah, 1991, 253). For Bellah, this Enlightenment myth is the cause of the "warfare" between religion and science—rather than anything inherent in religion or science. Bellah notes that in recent decades the open warfare between religion and science has given way to derisive indifference between the two parties—which is as detrimental to cultural well-being as open warfare. Bellah traces the origins of the progressive "desiccation of our culture" (Bellah, 1991, 244) to this rift between religion and science.

At the same time that Percy, Berger, and Bellah describe the different crises that we are facing, their work identifies some important resources that can help us to tackle the crises. Bellah, for his part, notes that while the dominant trend in the soft sciences has been to carry on aping the hard sciences and promoting Manichean metaphysics, a very different alternative is possible:

> The story I want to tell is that of another theory that also has its mythic dimensions, and that also has emerged out of social science itself, but that has a different conception of the human spirit, one in which religion has an integral place in a new conception of the unity of human consciousness. (Bellah, 1991, 237f.)

This alternative possibility is not a novel insight on the part of some late twentieth-century thinker. The work of "the most seminal minds in social science" from the late nineteenth and early twentieth centuries (i.e., Sigmund

Freud, Emile Durkheim, and Max Weber) is pregnant with this alternative possibility. In the work of "these three great non-believers" we find two apparently different trends: "Convinced of the invalidity of traditional religion, each rediscovered the power of the religious consciousness" 1991, Bellah, 240). Their work makes it virtually impossible to escape the conclusion that there is a dimension of empirical reality in being human and the social world that is composed of noncognitive, nonquantitative, and nonobjective elements. One cannot observe, study, or understand this dimension of this reality using the tools and concepts borrowed from the hard sciences any more than one can study biological reality using the tools and concepts borrowed from physics alone. In the decades that have passed since the days of Freud, Durkheim, and Weber, evidence has continued to accumulate that symbols that lie beyond the reach of the hard sciences "are constitutive of human personality and society [and] are real in the fullest sense of the word" (Bellah, 1991, 252). Based on his study of the early masters and the evidence that has accumulated since their time, Bellah says:

> I feel that there are greater resources now for healing the split between the imaginative and the cognitive, the intellectual and emotional, and the scientific and the religious aspects of our culture and our consciousness than there have been for centuries. Social science is beginning, faintly and crudely, to be able to cope with the richness of reality as religion has seen it. (Bellah, 1991, 245)

While Bellah sees the modern social sciences possessing untapped resources to heal the rift between religion and science, Percy offers the name of Charles Peirce as a resource who can help us to better understand the incoherence engendered in modern science by the "San Andreas Fault in the modern mind" (i.e., the mind/body dichotomy). Peirce's work is still relevant today because it not only sheds light "on the incoherence of science and of our own world view, but for its promise of contributing to a new and more coherent anthropology; that is, a theory of man" (Percy, 1991, 273). While Peirce made original discoveries in a number of different fields (most notably mathematics and geodesy), it is the one field that he founded (i.e., semiotics—the study of signs) that is most directly relevant to the issues at hand: "Peirce saw that the one way to get at it, the great modern rift between mind and matter, was the only place where they intersect, language" (Percy, 1991, 279). In short, Peirce's semiotics "laid the groundwork for a coherent science of man, and did so a hundred years ago" (Percy, 1991, 277).

The present inquiry takes the aforementioned insights of Percy, Berger, and Bellah as its point of departure. I very much agree with the diagnoses offered by the three regarding the crises in modern science, modern social science, and modern culture—these are three distinct but interrelated crises. Each of the three presents a daunting challenge in and of itself. All three taken together have brought the very legitimacy of modernity into question. At the beginning of the twenty-first century this legitimacy is being openly challenged by postmodernisms and fundamentalisms (both religious and secular) of different shades and varieties—and not just in the halls of academia but in politics, economics, art, religion, and every other sphere of culture. Since modernity is a human enterprise there are many dark spots and warts on its face. Being very much aware of these unattractive aspects, I nonetheless admire it for the contributions it has made to human culture at large. In my judgment some of the unique aspects of modernity are worth preserving and being passed on to future generations. I think the future will be the richer because of them and the poorer without them. At the same time I cannot close my eyes and ears to the deep angst caused by that which I admire—angst that expresses itself in some very pointed questions and critiques. To the degree that the questions by the relativists and fundamentalists are legitimate, they need to be addressed. And where the questions are not legitimate, it has to be demonstrated that such is the case to the satisfaction of the questioner. I will not make any pretense that this inquiry will address all the issues related to the three crises but I am confident it addresses some of the key issues. This confidence derives not so much from my admiration of my own intellectual prowess but from my reading and understanding of Weber and Peirce. One of the hallmarks of a classic is that it still speaks to us today especially in terms of helping us to understand ourselves better than we are able to do on our own. The works of Weber and Peirce are classics in this very sense. With these prefatory notes out of the way, we turn to a more detailed and technical part of the introduction.

Max Weber is recognized as a leading figure in the founding of the modern social sciences. While he is best known for his contributions in sociology, it is also well known that he wrote extensively on the methodology of social scientific inquiry. The relationship between his *Soziologie* and *Wissenschaftslehre* has been a matter of discussion and debate since Weber's own lifetime. Much of subsequent Weber scholarship has focused either on his sociology of culture or his methodology of social scientific inquiry. There are those who see Weber primarily as a sociologist of culture (i.e., Schluchter, 1979; Mitzman, 1970). Others see him primarily as a methodologist (i.e., Ringer, 2000). Bendix (1962) argues that there is no relationship between the two different aspects of Weber's work. While few others have gone as far as Bendix, the two aspects of Weber's scholarship rarely came into conversation with each other in Weber scholarship. In recent decades an increasing number of studies on Weber have gone against this trend (i.e., Whimster and Lash, 2006; Kalberg, 1996; Brubaker, 1984).

This relational reading of Weber has helped to resolve some of the longstanding controversies in the area of Weber studies. More importantly, the relational reading of Weber has shown that a number of Weber's insights make important contributions to some of the pressing contemporary cultural and academic debates. The present inquiry seeks to contribute to the trend of integrating Weber's sociological and methodological insights and demonstrating their continuing relevance. The main goal of this inquiry is to show that Weber's *Wissenschaftslehre* contains the resources to address the most significant modern cultural problematic identified by his *Soziologie* (i.e., the disenchantment of the world). In order to do an adequate job in this regard, some preparatory work is needed. The heated debates about Weber's *Wissenschaftslehre* are well known—mainly revolving around the issues of the ideal-type and value-free inquiry. Almost all of these debates focus on the technical details of Weber's methodological works—with the more general significance of his insights being obscured as a result. A number of Weber scholars have described the general significance of these insights, that is, Kalberg (1994), Hekman (1983, 1979), Huff (1984). From the perspective of the present inquiry, the most important of these insights is that Weber's *Wissenschaftslehre* contains the resources to integrate the *Naturwissenschaften* (the hard sciences) and *Geisteswissenschaften* (the soft sciences).[1] The basic reason why this significance has been overlooked is explicitly identified by Weber himself. Weber acknowledges that he does not have the philosophical resources (more specifically the logical and hermeneutical tools) to clearly articulate his understanding of the relationship between the hard and soft sciences. Most of the inquiry will focus on addressing this particular shortcoming. It will use the pragmaticism of Charles Peirce to bring clarity to certain aspects of Weber's *Wissenschaftslehre*. More specifically I will argue that Peirce's phenomenology, semiotics, and logic bring clarity to Weber's description of causality, the ideal-type, and the logic of scientific inquiry. Hence, when Weber's position is clarified using Peirce's pragmaticism, a relational vision of the hard and soft sciences almost forces itself upon the reader. This part of the inquiry will show that the Weber-Peirce synthesis is directly relevant to the crises in science and the soft sciences described above by Percy and Berger.

The foregoing suggests that this inquiry belongs to the areas of the philosophy of science and intellectual history. But from my perspective the primary value of this inquiry rests on showing that a relational reading of Weber's *Wissenschaftslehre* and *Soziologie* contains the resources to adequately understand and redress the most pressing modern cultural problematic—the disenchantment of the world. One of the places where this problematic manifests itself most dramatically is in the conflict between two spheres of cultures, the religious and intellectual. The fact that the Weber-Peirce synthesis can help to better understand and redress this problematic is hinted at by the following: Peirce's philosophy of religion is based on the same phenomenology, logic, and semiotics that are used to bring greater clarity to Weber's *Wissenschaftslehre*. When Weber's insights are viewed from the perspective of Peirce's pragmaticism, one gets a clear sense that Weber is also reaching for a relational vision of religious rationality and scientific rationality.

It is often assumed that Weber's understanding and practice of (social) science not only described the progressive rise of disenchantment and the decline of religion but also made a significant contribution to this very process. One of his contemporaries claimed that Weber was "mounting the most vitriolic of attacks upon religion" (Wittenberg, 1989 [1938], 118). This inquiry will argue that a relational reading of Weber's oeuvre shows that far from being an attack on religion of any kind, Weber's insights open up the possibility of a mutually enriching and affirmative relationship between religion and science by showing the deep affinity between religious rationality and scientific rationality. Different readers may find the present inquiry significant (or insignificant) for different reasons. For me, the significance (or insignificance) of the inquiry rests primarily (almost exclusively) on the following point: the ability of the inquiry to demonstrate that bringing Weber and Peirce into conversation reveals relationships between religion and science that are not otherwise apparent. This part of the inquiry will show that the Weber-Peirce synthesis is directly relevant to the crisis in modern culture (especially as it manifests itself in the religion vs. science divide) described by Bellah above.

CENTRAL THESIS

In the context of the *Methodenstreit* in late nineteenth-century and early twentieth-century Germany, Weber's *Wissenschaftslehre* made a significant contribution to formulating a relational vision of the *Naturwissenschaften* (the hard sciences) and the *Geisteswissenschaften* (the soft sciences). In very explicit terms, Weber notes that the difference between the hard and soft sciences "does *not* concern differences in the concept of causality, the significance of concept formation, or the kind of conceptual apparatus employed" (RK, 186). In other words, for Weber there is no difference between the hard and the soft sciences as far as: (1) using the notion of causality; (2) using concepts in scientific inquiry; and (3) the type of conceptual apparatus that is used. Consequently, at a basic level Weber rejects "the alleged difference between the 'logic' (or methodology) of the social and natural sciences" (Huff, 1984, 28). But this does not mean that Weber sees no difference between the hard and soft sciences. On the contrary, Weber argues that the two areas of scientific inquiry have certain unique, distinguishing characteristics. This distinction "concerns differences in (1) our causal *interest* and (2) the quality of 'self-evidence' pursued in the investigation of concrete causal relations" (RK, 186). In short, for Weber there is no difference between the hard and soft sciences at the formal, logical level, but there is an irreducible and fundamental difference between them at the qualitative and substantive level.

Weber's *Wissenschaftslehre* is positing the simultaneous unity and uniqueness of the hard and soft sciences—in one word he is offering a relational vision of the two branches of science. This allows Weber to go beyond the two opposing views of science that were dominant in his immediate intellectual milieu—the reductionism of the "positivists" and the dualism of the "intuitionists."[2] The positivists reduced the soft sciences to being merely a type of hard science and rejected the claim that there was anything unique about the soft sciences either in terms of their object of inquiry or the type of concepts used to study the objects during the course of inquiry. The intuitionists posited an unbridgeable epistemological or ontological divide between socio-cultural and physical-natural phenomena and rejected the claim that there was any similarity between the two. The significance of Weber's relational vision becomes clear when it is viewed in the context of the debates between the positivists and intuitionists. Against the positivists Weber argued that the soft sciences study aspects of empirical reality that the hard sciences cannot even conceptualize, let alone study. (This is similar to saying that the biological sciences study aspects of empirical reality that lie outside the domain of physics and that cannot be captured using the conceptual and methodological tools available to physics.) Hence the attempt to reduce the soft sciences into the terms of the hard sciences (in some type of "social physics" or the like) makes as much sense as the attempt to reduce the biological sciences into the terms of the physical sciences. At the same time Weber argued against the intuitionist claim that an unbridgeable ontological or epistemological divide separates the hard and soft sciences. Weber's *Wissenschaftslehre* avoids the reductionism of the positivists and the dualism of the intuitionists and offers a relational conception of the hard and soft sciences.

I will detail Weber's position using some of his earliest writings in the "methodology of the social sciences" (from the period 1903–1907). Some of these writings predate Weber's most well-known work, *The Protestant Ethic and the Spirit of Capitalism* (1904–1905), (i.e., "Rosher's 'Historical Method'" written in 1903). Others are contemporaneous with it (i.e., "'Objectivity' in Social Science and Social Policy," written in 1904, and part one of "Knies and the Problem of Irrationality" written in 1905), while still others followed shortly thereafter (i.e., part two of the Knies essay and "The Logic of the Cultural Sciences" in 1906 and *The Critique of Stammler* in 1907). Speaking of these essays, Toby Huff notes: "In these methodological writings, virtually all of the logical problems which continue to inspire controversy and debate in the social sciences were given paradigmatic formulation" (Huff, 1984,73). Huff describes the particular issues that Weber dealt with:

> We find Weber completely rethinking the methodology of the social sciences. To do so, he had to deal with all the fundamental questions about "laws" and "causality," prediction and alleged "irrationality" of human action, the difference between "self-evidence" and "validity," the nature of "understanding" and "interpretation" and of course aspects of the problem of the construction of "ideal-types." (Huff, 1984, 34)

In addition to containing "paradigmatic formulation" of many issues that continue to be debated in late twentieth- and early twenty-first-century social theory, Weber's work makes an irenic contribution to these debates.

In one such debate, Richard Peters and Peter Winch resurrected an old point of contention—the possibility of causal analysis in the social sciences. After describing the position of Peters and Winch as "the modern critique" of causal analysis in the social sciences, Susan Hekman argues: "Although Weber's concept of causality was formulated in response to another variant of the causality debate, elements of his position are particularly relevant to this new challenge" (Hekman, 1979, 67). Not only are Weber's insights relevant to the contemporary manifestation of the causality debate, "it can be maintained that Weber's concept of causality meets many of the modern arguments against the use of causal analysis in the social sciences" (Hekman, 1979, 76). In a similar vein, but on issues surrounding the ideal-type, Hekman notes:

> Weber's methodological position is undeniably a product of the methodological problems of his day and it grew out of discussions with his contemporaries as well as from questions arising from his own empirical work. Nevertheless, his theory is of more than merely historical interest. It is also applicable to the specific set of problems confronting contemporary social theory. (Hekman, 1983, 137)

In Hekman's view, since Weber's conception of the ideal-type "successfully synthesize[s] the analysis of subjective meaning and the assessment of structural forms," it is relevant to "a number of current problems in social theory" (Hekman, 1983, 120).

It is not just on isolated issues that Weber's *Wissenschaftslehre* continues to be relevant. Based on his reading of the methodology that is implicit in *The Protestant Ethic and the Spirit of Capitalism*, Stephen Kalberg argues that Weber's methodology is more advanced than all the dominant methodological positions that emerged in American sociology after WWII. The reason that Weber's position is superior to conflict theory, structural functionalism, modernization theory, comparative-historical sociology, the sociology of culture approach, and rational choice theory is that it avoids the dichotomies that characterize these other theories. From the perspective of his most well-known work:

> A particular view of the empirical world [emerges], one far distant from the major dichotomies that have structured the perception and understanding of this world among American theorists: history and sociology, conflict and equilibrium, society and the individual, culture and economic interest, social order and social change, individual interest and social structure, tradition and modernity and micro and macro levels of analysis. (Kalberg, 1996, 66)

Kalberg's assessment becomes significantly plausible in light of the following. A number of book-length studies put forth the argument that Weber's methodology has the resources to bridge dichotomies that continue to plague contemporary social theory—that is, Alexander (1982) on the ideal/real dichotomy, Ciaffa (1998) on the fact/value dichotomy and Ringer (2000) on the interpretation/explanation dichotomy.

Given the fact that Weber's writings remain a rich storehouse of "expansive theoretical capital" (Kalberg, 1996, 50) the question emerges as to why Weber's relational vision of science has not been adequately appreciated. Weber explicitly recognizes one reason himself—he lacks the resources he needs to articulate his relational conception of the sciences clearly. While he makes the following observation with respect to a very specific issue in the *Methodenstreit*, the observation is applicable more generally. Referring to the issue of the role of "empathy" and "reliving" in scientific inquiry as advocated by the intuitionists, Weber notes: "This question can only be answered by a *theory of 'interpretation,'* a theory which at this point is barely visible and has hardly been explored at all" (RK, 151). Weber makes this statement with sound knowledge of the different theories of interpretation that were in vogue in his intellectual milieu. This is evidenced by the note that he inserts at the end of the fore cited observation:

> The works of Schleiermacher and Boekh on "hermeneutics" are irrelevant here since they do not pursue epistemological aims. The account by Dilthey contained in the *Proceedings of the Berlin Academy* (1894) and decisively rejected by psychologists (Ebbinghaus) is weakened by the following prejudice: the belief that specific systematic *sciences* must correspond to certain formal categories of knowledge. (RK, 250, n.46)

The following observation by Weber gives us a clue regarding the type of theory he is looking for: "Sociology must reject the assumption that 'understanding' and causal 'explanation' have *no* relationship to one another" (CIS, 157). In short, Weber needs a theory of interpretation that produces a relational conception of "understanding" and "explanation."

In addition to explicitly acknowledging that the available theories of interpretation are inadequate for his needs, Weber's writings contain implicit pointers that the available theories of the logicians are also inadequate. Weber acknowledges that he is relying on the work of the modern logicians in formulating his methodology. In the introductory section of "'Objectivity' in Social Science and Social Policy" he notes: "Those who know the work of the modern logicians—I cite only Windelband, Simmel and for our purposes particularly Heinrich Rickert—will immediately notice that everything of importance in this essay is bound up with their work" (OSS, 50). There is significant consensus in Weber scholarship that even though Weber is using the work of contemporary logicians (especially Rickert) there is little doubt that he is going beyond them.[3] The basic characteristic of the type of logic that Weber needs is that this logic makes *Sinn* (meaning) an objective possibility in the universe and at the same time allows for the scientific investigation of *Sinn*. By acknowledging the role of values in socio-cultural inquiry, Rickert's logic comes closest to Weber's needs. But it still falls short because Rickert's theory forestalls the logical analysis of the validity of these values. In light of what Weber has acknowledged, the following observation by Pierre Bourdieu is very much on the mark: "So far as we can see, the intellectual tools Weber had at his disposal prevented him from forming a clear awareness of the principles he was applying (at least intermittently) in his research and consequently from setting them to work in a methodical and systematic fashion" (Bourdieu, 1987, 121). Because Weber lacks the adequate intellectual tools to articulate his relational vision of science clearly, there is a lack of clarity in Weber's *Wissenschaftslehre*. This lack of clarity is partially responsible for the appearance of inconsistencies in his work and many of the controversies that these apparent contradictions have given rise to in subsequent Weber scholarship.

Weber notes that the single most important function that we can demand of science is that it brings greater conceptual clarity and understanding to our ideas and actions (SV, 150–152). Clarification is precisely the relationship that is being proposed between Weber and Peirce in the present work. In exploring this possibility, this inquiry is following through on some insightful and tantalizing hints offered by Huff (1984).[4] Certain elements of Peirce's pragmaticism clarify certain elements of Weber's *Wissenschaftslehre*. More specifically Peirce's phenomenology, semiotics, and logic clarify Weber's position on the three issues that he identifies as the shared formal, logical ground of the hard and soft sciences: (1) causality, (2) concept formation, and (3) conceptual apparatus. This is particularly helpful because Peirce's work is still acknowledged as a resource in contemporary philosophy of science. This inquiry will provide added evidence in this regard by showing that some of the key moves made by Weber in his *Wissenschaftslehre* are similar to the moves made by Peirce in his pragmaticism. This part of the inquiry will help us to better understand the contemporary manifestation of the crises within the sciences as described by Percy and Berger.

After reconstructing Weber's *Wissenschaftslehre* using Peirce's pragmaticism, the inquiry will turn to the cultural crises engendered by disenchantment as described by Weber's sociology. Weber's sociological inquiry shows that the progressive development of science has been "a fraction, the most important fraction" (SV, 138) of the process of disenchantment that has been unfolding for millennia. In modern times this process reaches its climax, bringing with it intellectual and practical challenges that are unique to this period of human history. In intellectual terms it means the death of *Sinn* (meaning) in the universe; in practical terms it means inevitable and irreconcilable conflict between the different spheres of culture. Weber's investigation reveals that while the origins of these crises go back to the fragmentation and differentiation of scientific rationality from religious rationality, the most egregious manifestation of the crises takes place when the two rationalities become completely independent and autonomous in modern times. The inquiry will show that just as Weber's *Wissenschaftslehre* allows us to go beyond the reductionist and dualistic conceptions of science and envision a relationship between the hard and soft sciences, his insights that science is a cultural activity allow us to beyond the reductionist and dualist conceptions of scientific rationality and religious rationality and reveals a deep affinity between the two.[5] The inquiry will conclude by demonstrating that just as Peirce's pragmaticism helps us to clarify Weber's position in terms of the methodology of scientific inquiry and see the intimate relationship between the hard and soft sciences, Peirce's insights transform the Weberian deep affinity between scientific and religious rationality into a mutually enriching relationship between religion and science.

CHAPTER SUMMARIES

The bulk of the inquiry will show that a relational conception of the hard and soft sciences is implicit in Weber's work by looking at the way causality, concept formation, and conceptual apparatus are employed in scientific inquiry. Chapter 1 will present Weber's position on causality in the soft sciences, using the causal account of the origin of the spirit of capitalism as an example. The chapter will also describe Weber's critique of the positivist and intuitionist positions on causality. Weber's critique of the positivists shows that while one has to keep certain rules and laws in mind when determining the cause of any phenomenon, the rule or law cannot be accorded causal agency. In his critique of the intuitionists, Weber shows that while human freedom and creativity can be causal agents under certain circumstances, this does not mean that "mechanical" natural processes can never be causal agents. Weber's analysis demonstrates that just as it is wrong to accord causal agency to abstract rules and laws, it is wrong to attribute causal agency to something that is uniquely human. Whereas chapter 1 will describe Weber's conception of causality in light of his critique of the positivist and intuitionist conceptions, chapter 2 will reconstruct Weber's position on causality. The main goal of this chapter is to detail Weber's argument that while the hard sciences can employ the concept of causality only in its truncated and degenerate form, the soft sciences "employ the category of causality in its full meaning" (RK, 196).

Chapter 3 will describe Weber's position on concept formation. The concept is embedded in the very definition of science offered by Weber. Consequently, all scientific inquiry involves concepts—but the type of concept that is used differs according to the type of inquiry that is being carried out. If one is engaged in mathematical inquiry one uses concepts that are different from the types of concepts that are used in the study of plants and animals. If one is studying planets and black holes, one uses concepts that are different from those used in biology. Similarly the concepts employed by the soft sciences are different from the concepts employed by the hard sciences and just as the invention of new concepts (let's say by biology) laid bare aspects of empirical reality that remained hidden from physics, the use of concepts that are particular to the soft sciences opens up aspects of empirical reality that remain beyond the reach of the hard sciences. Chapter 3 will detail this point and describe how the ideal-type is the typical concept used by the soft sciences. The ideal-type makes it possible to conceptualize qualities, values, and beliefs, thereby making it possible not only to see the aesthetic and ethical dimensions of empirical reality but also subject them to scientific scrutiny. This chapter will describe Weber's critique of the positivist position that sees the soft sciences using the same type of concepts as the hard sciences. The critique of the positivist position will be complemented by Weber's critique of the intuitionist position that there is no need for the use of concepts in socio-cultural inquiry. A key argument in chapter 3 will be that Weber's critique of the intuitionists contains a typology of experience that is laid bare and clarified by Peirce's phenomenology. Chapter 4 will reconstruct Weber's position on concept formation. This reconstruction will show that at the formal, logical level there is no difference between the ideal-type (the type of concept employed by the soft sciences) and the generic concept (the type of concept employed by the hard sciences). At the same time, there is an irreducible, substantive difference between the two. The ideal-type makes possible the conceptual mastery of empirical phenomena in terms of their uniqueness/individuality, change and development over time, and their cultural meaning and significance. In contrast the generic concept makes possible the conceptual mastery of empirical phenomena in terms of their universality, immutability, and from the perspective of impersonal, detached objectivity. This chapter will extract a typology of concept from Weber's work and then look at this typology from the perspective of Peirce's semiotics. In terms of Peirce's semiotics, Weber's ideal-type and generic concept are both symbols—but the ideal-type is a particular type of symbol (a genuine symbol) while the generic concept is a degenerate symbol. In addition to helping us better appreciate Weber's insights, Peirce's semiotics shows that while the hard sciences can only employ the concept in its truncated, degenerate form, the soft sciences employ the concept "in its full meaning." Clarifying Weber's position on concept formation using Peirce's phenomenology and semiotics helps us to see that Weber's *Wissenschaftslehre* expands the scope and horizons of scientific inquiry by laying bare dimensions of empirical reality (i.e., aesthetics and ethics) that remain beyond the reach of the hard sciences. At the same time it makes it possible to subject the characteristics of the phenomena (i.e., the unique individual, variation/change, and significance/meaning) in these dimensions to scientific inquiry.

Chapter 5 will move on to the topic of conceptual apparatus. Weber argues that the "apparatus" used in scientific inquiry is composed of: (1) intuitive insight, (2) abstraction and generalization, and (3) comparison and contrast. What Weber calls "conceptual apparatus" is actually the logic of scientific inquiry. In terms of Peirce's logic, scientific inquiry is composed of abduction, deduction, and induction. When clarified by Peirce's logic, Weber's description of conceptual apparatus not only establishes the "objective" reality of "subjective" elements like qualities, values, and beliefs, it makes an investigation of these elements as much of a scientific undertaking as inquiry into atoms, chemicals, and photosynthesis. The second half of the chapter will translate Weber's reflections on conceptual apparatus into the hermeneutics that emerges from Peirce's semiotics and logic. By the end of the chapter we will be able to see that what Peirce labels class-8, -9, and -10 signs helps us to not only better understand Weber's insights on the relationship between understanding, explanation, and interpretation, but also gives Weber the type of "theory of 'interpretation'" that he needs. This theory on the one hand lays bare the role of the investigator in shaping scientific inquiry and on the other hand makes it possible to ask certain types of investigative questions that are not otherwise possible. It becomes possible to investigate the meaning, cause, and reason of events in empirical reality and show how these elements cannot be separated from the investigative concerns of the inquirer. In line with what has been said about causality and concept formation, Weber shows that the soft sciences employ conceptual apparatus in its "full meaning," while the hard sciences cannot but employ it in its truncated, degenerate form. The hard sciences can neither acknowledge the role of abduction in scientific inquiry, nor subject the abductive elements in scientific inquiry to logical scrutiny.

The fact that the soft sciences employ scientific logic in its "full meaning" can be demonstrated by looking at the key point on which the hard and soft sciences differ in terms of their ultimate goal. With respect to the empirical reality that is the object of scientific inquiry and of which scientific inquiry is a constitutive part, we can say the following about the difference between the hard and soft sciences. Weber's *Wissenschaftslehre* (clarified by Peirce's pragmaticism) shows that scientific inquiry (like any other cultural activity) is not possible in the absence of certain subjective elements. Scientific investigation of empirical reality forever remains truncated and degenerate if the subjective elements that are part of the cultural activity called "science" are not laid bare and subject to objective analysis—this is a task that the hard sciences can neither conceptualize nor carry out. In making possible the objective analysis of the subjective elements in science, the soft sciences further expand the horizons of scientific inquiry by opening up those dimensions of empirical reality to scientific scrutiny that remain beyond the reach of the hard sciences. In more precise terms, Weber's logic of scientific inquiry reveals certain elements in the process of scientific inquiry that the inductive and/or deductive philosophies of science cannot recognize and still less subject to scientific scrutiny.

There are practical and concrete implications of moving from a truncated, degenerate conception of scientific inquiry being composed of only inductive and/or deductive elements and recognizing its triadic character. In addition to expanding the breadth of scientific inquiry, the soft sciences increase its depth by making it possible to ask certain types of investigative questions that are neither theoretically nor logically possible from the perspective of the (inductively-deductively conceived) hard sciences. In light of the discussion in the previous four chapters the most important of these questions is: "Why are events in empirical reality as they are when they could have been otherwise?" For Weber it is only when this investigative question is asked and a testable hypothesis offered as an answer that the event(s) under investigation become "rational." To answer this question in terms of the "laws of nature," or "something inherent in human nature" is to not just beg the question but to give a patently wrong, logically absurd, and scientifically meaningless answer. Even if these types of answers are accepted, the next question would be: "Why are the 'laws of nature' (or 'human nature') as they are when they could be otherwise?" In the absence of a testable hypothesis as an answer to this question, the inquiry terminates at a cosmic riddle or is faced by an infinite regress. Such a conclusion to a discussion may be acceptable in certain religious theologies (even that is a very long stretch) but this conclusion cannot be accepted in scientific inquiry under any circumstances (as "science" is conceived by Weber and Peirce). In short, Weber's *Wissenschaftslehre* expands the horizons of scientific inquiry by demonstrating that empirical reality has aesthetic and ethical dimensions and is characterized by individuality, change, and significance—all the while bringing these dimensions and characteristics under the gaze of scientific scrutiny. At the same time, Weber deepens the reach of scientific inquiry by making it possible to ask certain types of investigative questions that are not otherwise possible—most notably the question: Why is X as it is when it could be otherwise?

Having clarified Weber's *Wissenschaftslehre* using Peirce's pragmaticism in chapters 1 through 5, chapter 6 will explore its significance from the perspective of a key finding in his *Soziologie*, that is, that disenchantment is the "fate of our times" (SV, 155). In terms of theoretical rationality, disenchantment means the death of *Sinn* (meaning) in the universe. In terms of practical rationality, disenchantment means perpetual and irreconcilable conflict between the different spheres of culture. Weber's sociology of culture identifies the progress of science as the most significant causal factor in the growth, development, and spread of disenchantment. Weber's *Wissenschaftslehre* helps us to see that science *per se* is not the cause of disenchantment but rather a very particular conception of science. In more precise terms we can say that the truncated, degenerate conception of science is the cause of the disenchantment of the world. The fact that Weber offers a fuller and more genuine conception of science as an alternative opens up the possibility that science can have a different relationship with culture than disenchantment. The discussion will begin by showing that the disenchanting conception of science and Weber's relational alternative are both the products of the same process, that is, the progress of science. Then the discussion will demonstrate that the particular understanding of causality (natural causality), concept formation (generic concept), and conceptual apparatus (deduction-induction) that is characteristic of the degenerate conception of science leads to the disenchantment of the world. This understanding of causality, concept formation, and logic not only leads to an unbridgeable divide between the hard and soft sciences but also an inevitable and irreconcilable conflict between the intellectual and religious spheres of culture—or any chosen two spheres of culture for that matter.

The discussion will then look at how the different understanding of causality, concept formation, and logic that characterizes Weber's relational conception of science radically changes the relationship between science and the nonscientific spheres of culture (more specifically, the one sphere that is furthest removed from science, i.e., religion). When we combine Weber's reflections on the sociology of culture and religion with his relational conception of science we begin to notice points of affinity between the religious sphere and the intellectual sphere (in the form of science). The first point of affinity is that when viewed as cultural activities, religion and science are two different interpretations of the same set of presuppositions: (1) Meaning is real, (2) Meaning can be had, and (3) Meaning is worth having. The second point of affinity that Weber's analysis reveals is that there is no such thing as science without the presence of certain subjective elements such as the affirmation of faith, passionate devotion, and the gift of grace—subjective elements that can only be labeled "religious." The third point of affinity is the fact that from Weber's perspective, historians and sociologists know of no religion in human history that was devoid of certain objective elements such as rational coherence, self-critical objectivity, and worldly success—elements that are glaringly "scientific." Each of these points evidences that there is affinity between the religious and scientific spheres of culture. The three points taken together evidence a deep affinity between the two. The deep affinity between scientific rationality and religious rationality that is implicit in Weber's relational conception of science, in its turn, opens up cultural possibilities that cannot even be imagined from the perspective of the degenerate, truncated conception of science—the possibility of establishing relationships between elements and spheres of culture rent asunder and condemned to perpetual, irreconcilable conflict. The cultural significance of Weber's *Wissenschaftslehre* lies in the fact that it opens up cultural horizons by pointing toward affinities and possible relationships that are not otherwise apparent.

Chapter 7 will show that there is nothing accidental in the discovery of deep affinity between religion and science that is implicit in Weber's work. Such an affinity is to be almost expected given the fact that Peirce's philosophy of religion is based on the same phenomenology, logic, and semiotics that clarified Weber's reflections on the methodology of scientific inquiry. As a matter of fact, different parts of Peirce's work not only affirm the three points of affinity between religion and science implicit in Weber's work but transform the affinity into a mutually enriching relationship. In his "A Neglected Argument for the Reality of God" Peirce notes that the relationship between his theory of scientific inquiry and his argument for the reality of God is so intimate that if the former is accurate the latter is true—both of them being based on the exact same phenomenology, logic, and semiotics. This affirms Weber's insight that religion and science are different interpretations of the same set of presuppositions. After noting that scientific logic presupposes the validity of inferential reasoning, Peirce goes on to explore the grounds of validity of this logic. In his "The Doctrine of Chances" Peirce argues that inferential reasoning is not possible in the absence of three sentiments—faith, hope, and love—which are, strictly speaking, religious sentiments. This deepens Weber's insight that scientific inquiry is made possible by not just any sentiments but by those (as Peirce notes) identified by the Bible as the most worthy of cultivation. A close reading of Peirce's essay "A Religion of Science" shows that, in addition to its other functions, religion has a very specific worldly function. Between the lines Peirce is arguing that science requires the emergence of a particular type of religious attitude if scientific knowledge is to have any meaning or significance. In the absence of this religious attitude, all the triumphs of science amount to almost nothing.

The Weber-Peirce synthesis leads to the following conclusion: A scientifically valid understanding of science requires rejecting the reductionist and dualistic conceptions of the hard and soft sciences and replacing them with a relational understanding. The relational understanding of science logically necessitates rejecting the reductionist and dualistic conceptions of religion and science and replacing them with a relational understanding. This relational understanding of religion and science is a part of (and perhaps a prelude to) bringing not only the intellectual and religious spheres of culture into relationship but also bringing the political, economic, aesthetic, erotic spheres into relationship with each other as well as the intellectual and religious spheres. Expressed in terms used in the opening pages of the introduction, the constellation of historical factors that made possible the emergence of modern Western culture has indeed been transformed significantly over the past three centuries. This transformation has in its turn curtailed the range of freedom, choice, and autonomy of the cultural beings who are a part of the constellation. While a superficial view of the cultural condition suggests that there is no alternative to the "iron cage" and that a resigned acceptance of disenchantment as "the fate of our times" is the only "logical," "rational," and even "scientific" option, a relational reading of Weber and Peirce points toward a very different alternative. The fact that both Weber and Peirce are located at the core of the very same historical constellation that gave rise to the modern disenchanted cultural condition points to not only the logical, rational, and scientific character of the relational alternative but also to its validity and significance.

ADDENDUM TO THE INTRODUCTION

The *Methodenstreit*: A Brief Overview

Weber's reflections on the philosophical and methodological issues in scientific inquiry took place in a particular historical context—during the late nineteenth and early twentieth centuries. It might be helpful to describe some of the important features of this context. With the emergence of new disciplines such as economics, history, and sociology, the term "science" came under increasing scrutiny. In order to establish their "scientific" credentials, each of these new disciplines needed clarity as to what the term "science" actually meant. If the meaning (i.e., definition) of the term was already fixed then the emergent disciplines had to conform to this meaning. If the meaning was not fixed then the question emerges: What is the meaning of the term? Generally speaking, the default position in the discussion was that the meaning of "science" was already known and the emergent disciplines had to conform to this meaning. In hindsight, it is obvious that this default position would cause some deep and intractable problems—and it did cause such problems and gave rise to what came to be called the *Methodenstreit* (the battle of the methods). This "battle" was originally a debate involving two leading economists in the 1890s, Carl Menger and Gustav von Schmoller.

Menger advocated an "abstract" approach in economics, and von Schomoller an "empirical" approach. While originating in the area of economics and originally focusing on the general theme of deduction vs. induction, the debate spilled over into other areas and began to touch upon other issues. When other parties and other themes became a part of the debate, the key question that emerged in the battle was: "Can a discipline whose methods illuminate the physical and biological world be applied to the sphere of human action?" (Diggins, 1996, 114). Leaving aside many important details, we can identify four major positions in this "battle"—the Classical School of Economics, the Historical School of Economics, the *Verstehen* School, and the neo-Kantian School. A stick figure of each of the four positions is as follows.

The Classical School of Economics

It viewed social reality the same way that Newtonian physics viewed natural reality. While the events in observed reality are multiple and constantly in flux, the overall structure of reality is fixed and static. Underpinning the observed multiplicity, flux, and individuality of events there are immutable, universal, and impersonal laws. Consequently, from this perspective the method of socio-cultural inquiry is no different from that of the hard sciences (more specifically from Newtonian physics)—"to discover in the multiplicity of events the lawlike rule of simple forces" (RK, 58). For this position there was no difference between the hard and soft sciences either in terms of the "objects" that were being studied or in the methods used to study the "objects."

The Historical School of Economics

Wilhelm Roscher, Karl Knies, and Bruno Hildebrand are considered the founders of this school. In contrast to the Newtonian view of the physical universe, this position is best described by the analogy of the Darwinian view of the biological world. Looking at social reality it is obvious that there is dynamism, change, and development (one may even say "evolution") in social reality. Like Darwin, the proponents of the Historical School sought to discover the "laws" that determine the course of historical change. While Weber counts himself among the "offspring" of the Historical School (OSS, 106), he categorically rejects the goal of socio-cultural inquiry as envisioned by this school. Describing the goal of the Historical School, Weber notes:

> The latter still hold in many ways, expressly or tacitly, to the opinion that it is the end and the goal of every science to order its data into a system of concepts, the content of which is to be acquired and slowly perfected through observation of empirical regularities, the construction of hypotheses and their verification, until finally a "completed" and hence deductive science emerges.

> For this goal, the historical-inductive work of the present-day is a preliminary task necessitated by the imperfection of our discipline. (OSS, 106)

For this school, while there is a difference in the "objects" of the physical-natural world and the socio-cultural world, the methods by which the "objects" are to be studied are the same. In terms of philosophical resources, "it was perhaps John Stuart Mill who had provided the clearest and most uncompromising statement" (Oakes, 1975, 21) of the position advocated by the Historical School. In the terms used by Hekman and Oakes, the Historical School was the proponent of the "positivist" position.

The Verstehen School

The most well-known proponent of this position was Wilhelm Dilthey. From the perspective of this school there is something ontologically unique about the socio-cultural domain—this is the domain of *Geist*. The physical-natural domain is completely devoid of *Geist*. In observing physical-natural phenomena, the knowledge of the observer is always incomplete and imprecise because in obtaining knowledge of a domain that is devoid of *Geist*, the *Geist* of the observer remains passive. In contrast, when studying socio-cultural reality the observer can attain intuitive and infallible knowledge because he can empathize with or relive the experience of the actor who is being observed. This intuitive knowledge allows the investigator to go beyond mere "explanation" of the actor's behavior to actually "understand" the behavior. In its early years this position was influenced by the phenomenology of Franz Brentano and Edmund Husserl. For the *Verstehen* School there is an epistemological and ontological divide between the hard and soft sciences. It is by positing this absolute divide that the *Verstehen* School attempted to maintain the autonomy of the soft sciences. While Weber did adopt the notion of *Verstehen* (or "understanding") from this school, he rejected their methodological and metaphysical reflections on the nature and role of *Geist* in scientific inquiry. In the terms of Hekman and Oakes, the *Verstehen* School is representative of the "intuitionist" position.

The neo-Kantian School

While the efforts of the *Verstehen* School to challenge the reductionist position of the Classical and Historical Schools were admirable, their dualistic solution to the problem was found to be wanting by the neo-Kantian School: "Weber's neo-Kantian contemporaries, reacting to a lofty Hegelian metaphysics that made no effort to assimilate natural science, sought to find ways to use philosophy in order to have access to truths about history, society and culture" (Diggins, 1996, 114). Wilhelm Windelband and his younger protege Heinrich Rickert were the neo-Kantians who most directly influenced Weber. Instead of focusing on the "object" of inquiry or the "tools" used to study the "object," the neo-Kantians focused on the role of the investigator in the inquiry. Windelband noted that the socio-cultural inquirer is interested in phenomena because of their uniqueness, while the physical-natural scientist in interested in phenomena because of their uniformity. It is this difference in the orientation of the inquirer that distinguishes the soft sciences from the hard sciences. Windelband's insight allows us to see that "for the purpose of scientific analysis, any given lump of coal would be considered interchangeable with other individuals of its type, whereas Goethe would not be similarly interchangeable with other poets or historical figures of his age" (Ciaffa, 1998, 53). In short, the hard sciences pursue "nomothetic" knowledge (knowledge of generals), while the soft sciences pursue "ideographic" knowledge (knowledge of individuals). Rickert builds on Windelband's insights, especially with regard to the role of investigator studying individuals. Rickert noted that even when an investigator is looking at a unique individual she is looking at an "illimitable manifold" and in itself the unique individual is "immeasurable" and "cannot be grasped as it is" (quoted by Ciaffa, 1998, 46). The investigator plays a critical role in determining which segments of the "immeasurable" parts of the individual will become the focus of scientific inquiry. In making the selection the values of the investigator play a decisive role. From among his contemporaries, this is the position that Weber was closest to. Rickert offered Weber a theory of the "illimitable manifold" character of empirical reality, the unique individual as an "illimitable manifold," and the role of the investigator's values in making this manifold manageable. But even Rickert did not give Weber a theory of interpretation that could bridge the gap between understanding and explanation and make scientific investigation of *Sinn* (meaning) possible. It is from among Weber's reflections, criticisms, and insights that emerged in the midst of this *Methodenstreit* that we will gather pieces of Weber's *Wissenschaftslehre* and put them together into a coherent whole.

NOTES

1. These terms have been translated differently in Weber scholarship. The terms have a history prior to Weber and a history after him. For the purposes of this inquiry I will use the most economical translation of these terms, i.e., the hard sciences and the soft sciences.

2. Here I am employing the terminology used by Susan Hekman (1979) in "Weber's Concept of Causality and the Modern Critique" and by Guy Oakes (1975) in his "Introductory Essay" to Weber's *Roscher and Knies: The Logical Problems of Historical Economics*. Hekman notes:

> In the monographs in which he treats the problem of causality Weber is concerned to dispose of two doctrines that constituted the polar position of the current debate over the methodology of the social sciences, the Methodenstreit. The first position, taken by the positivists, asserts that the same form of causal analysis is appropriate

> to both the natural and social sciences. The second, espoused by the intuitionists, argues that a metaphysical belief in the freedom of the will excludes the application of abstract concepts and generalizations to the description of human conduct. (Hekman, 1979, 68)

Guy Oakes (1975) also identifies positivism and intuitionism as the two poles in the *Methodenstreit* that Weber "attacks" (Oakes, 25–32).

3. Guy Oakes (1988) has made the most forceful claim that Weber does not go beyond the logic of Rickert. This position has been analyzed and soundly critiqued by Ringer (2000) and Huff (1984), among others.

4. The original source of this insight on my part was Huff (1984). In his short, insightful study of Weber's methodology, Huff identifies three particular points on which the similarity between the two thinkers is striking. Taking the similarity on these three as the starting point, I began to explore the possibility that the similarity between the two extends beyond just a few isolated points.

5. This broadens and deepens a claim I have made earlier that a relational reading of Weber's methodological and sociological writings contains the resources to "disenchant disenchantment." This position was initially put forth in Koshul (2005) *The Postmodern Significance of Max Weber's Legacy*. The present inquiry sees Weber not merely addressing (or disenchanting) disenchantment but the Weber-Peirce synthesis containing the resources to redress it.

Chapter One

Causality and Scientific Inquiry

Our starting point for constructing Weber's relational conception of science is his observation that the difference between the soft and hard sciences "does *not* concern differences in the concept of causality, the significance of concept formation or the kind of conceptual apparatus employed" (RK, 186). This insight will be combined with Weber's observation that there is a clear difference between the soft and hard sciences that is determined by the causal interest of the investigator and the type of evidence that the investigation aims at. During the course of our inquiry we will take a close look at Weber's critique of the positivist and intuitionist conceptions of causality, concept formation, and conceptual apparatus. This review of the positions that Weber critiqued will show that one cannot even conceive of the possibility of a relationship between the soft and hard sciences from the positivist and intuitionist perspectives. The positivist position produces a reductionist conception of science where the soft sciences are reduced to being a special case of the hard sciences. The logic underpinning this move is the same as the logic that reduces psychology to a special case of biology and then further reduces biology to a special case of physics. The intuitionist position produces a dualistic conception of science where an abyss separates socio-cultural reality from physical-natural reality—with the former being interpretable or understandable in a way that the latter can never be. As we look at Weber's analysis of the logical and methodological errors that lead to positivist reductionism and intuitionist dualism, we will see that an outline of his own relational alternative in implicit in the critique.

Beginning with Weber's definition of the soft sciences, we will detail his argument on the role causality, concept formation, and conceptual apparatus in scientific inquiry. For Weber the soft sciences are a particular type of "empirical sciences of concrete reality"—he describes the latter in these terms:

> The type of social science in which we are interested is an *empirical science* of concrete *reality* (*Wirklichkeitswissenschaft*). Our aim is the understanding of the characteristic uniqueness of the reality in which we move. We wish to understand on the one hand the relationships and the cultural significance of individual events in their contemporary manifestations and on the other the causes of their being historically *so* and not *otherwise*. (OSS, 72)

Without losing sight of the earlier observation about what the difference between the soft and hard sciences does not entail, this passage identifies the distinguishing features of the soft sciences that set them apart from the hard sciences. Firstly, the soft sciences aspire to understand the "uniqueness of the reality in which we move." Secondly, they aspire to understand the "cultural significance of individual events" for the individuals experiencing those events. Thirdly, they aspire to understand the cause of this uniqueness and the significance of the event being as it is when it could have been otherwise. In short the soft sciences aspire to attain scientifically objective knowledge about unique individuals, cultural significance, and the cause of events being as they are when they could have been otherwise. In their pursuit of this knowledge the soft sciences employ the categories of causality, concept formation, and conceptual apparatus at least as much as the hard sciences.

The difference between the hard and soft sciences is not at the "formal, logical" level but at the qualitative, substantive level. What Weber means by the terms "formal, logical" and "substantive" can be understood in light of his description of "formal rationality" and "substantive rationality." Kalberg notes that Weber uses the term "formal rationality" to mean the pursuit of consistency in thought or action "by reference back to universally applied rules, laws, or regulations" (Kalberg, 1980, 1158). In contrast, substantive rationality refers to the pursuit of consistency in thought or action "in relation to past, present, or potential 'value postulates'" (Kalberg, 1980, 1155). Consequently, there is no difference between the hard and soft sciences if the categories of causality, concept formation, and conceptual apparatus are understood from the perspective of a "universally valid" and "objective" definition. But there is a difference between the two branches of scientific inquiry if these three categories are understood from the perspective of the particular "value postulates" that shape the two types of scientific inquiries. For Weber the qualitative difference between the hard and soft sciences does not compromise the scientific status of the soft sciences—quite the contrary. The unique characteristics of the soft sciences make an invaluable contribution to our understanding of science by laying bare critical aspects of scientific inquiry that remain hidden, unaccounted for, or otherwise taken for granted by the hard sciences.

With specific reference to the aforementioned three categories, the emergence of the soft sciences in relationship to our understanding of science is very much like the relationship of the discoveries of Einstein and Heisenberg to our understanding of physics. Einstein and Heisenberg did not prove Newtonian physics to be "wrong," they demonstrated the limitations of the definition of "physics" emerging from Newtonian physics. This definition was largely based on the understanding of the "laws of physics" discovered by Newton. Relativity and quantum physics laid bare aspects of empirical reality where the "laws of physics" were not just different from the Newtonian "laws of physics" but actually the very opposite. Similarly, the manner in which the soft sciences employ the categories of causality, concept formation, and conceptual apparatuses does not show that the manner in which the hard sciences employ them is "wrong," but it does conclusively demonstrate that it is severely limited. These three categories are far more than the hard sciences are capable of imagining (in the same sense that matter, cosmos, and universe are far more than Newtonian physics is capable of imagining). If this is indeed the case then the implications of Weber's conception of the soft sciences is as significant and revolutionary for science's self-understanding as the discoveries of Einstein and Heisenberg were for physics' self-understanding. The implications are that the emergence and development of the soft sciences contribute to a more nuanced, well-rounded, and comprehensive understanding of science—in short, the soft sciences contribute to a more scientific understanding of science than the one that the hard sciences are able to offer on their own. Weber opens the way to this more scientific understanding of science by going beyond the reductionism of the positivists and the dualism of the intuitionists and offering a relational conception of the soft and hard sciences.

We will begin the exposition of Weber's relational conception of science by looking at the formal unity and substantive uniqueness of the category of causality in the soft and hard sciences. Weber argues that science is one way of knowing the world among others—but it is a unique way of knowing. One of the unique characteristics of scientific knowledge is that it offers an "interpretation" of empirical reality—but "interpretation" in a very specific sense. Weber notes: "Whenever 'interpretation' is employed in empirical *science*, *then* we find it in the form with which we are concerned in this essay . . . it is a form of *causal* knowledge" (RK, 154f.). Weber emphasizes, again and again, that it is crucial to understand the unique characteristic of scientific interpretation as a type of "*causal* knowledge" because it is precisely this quality that distinguishes scientific knowledge from other types of knowledge. For Weber, interpretation means "the imputation of a judgment, in the sense of an affirmation that an *actual* set of relations is validly 'understood.' This is the only sort of 'interpretation' we are discussing here, the sort of 'interpretation' which produces knowledge of *causal* relations" (RK, 149).

For Weber, sciences of empirical reality are (or at least should be) solely concerned with "producing knowledge of *causal* relations" or causal interpretation.

It is clear that Weber is aware of the fact that "interpretation" can be defined differently than he has used the terms in the phrase "causal interpretation"—and that the different definition is also valid. As a matter of fact he discusses other types of interpretation, repeatedly, to emphasize the fact that scientific inquiry should remain solely committed to the pursuit of causal interpretation—and it can only do this by heightened awareness of the types of interpretations that it should avoid. Artistic interpretation and teleological interpretation are two examples of noncausal interpretation. Weber notes that in addition to the imputation of causal relations, interpretation can be "the imputation of a '*valuation*' of a specific quality" (RK, 149). Such an interpretation would be "an allusion to a specific emotional *commitment*—for example, the 'suggestiveness' of a work of art or of the 'beauty of nature'" (RK, 149). This type of interpretation is the defining characteristic and ultimate goal of aesthetics. In addition to causal interpretation and artistic interpretation there is the possibility of still another type of interpretation—teleological interpretation. In this form interpretation means the imputation of the "validity" of particular means for the purpose of attaining a predetermined end. This type of interpretation is the defining characteristic of ethics. Weber argues that the superiority of one type of interpretation over the other types is a matter that cannot be decided scientifically. Choosing one over the others depends on the type of knowledge that the individual is seeking. What he does say is that causal interpretation offers a unique type of knowledge that cannot be had from any other source—not from artistic interpretation (aesthetics), from teleological interpretation (ethics), or interpretation of any other type. Those who value causal interpretation (for whatever reason) must turn to science. (Weber's reasons for valuing causal interpretation will emerge during the course of the discussion.)This chapter will begin with an example of a causal account offered by Weber and then describe Weber's critique of the positivist and intuitionist understandings of causality. The purpose of this discussion is to gather the pieces that will be used to reconstruct a coherent account of Weber's conception of causality (in the next chapter).

A CAUSAL ACCOUNT OF THE ORIGIN OF CAPITALISM

Instead of beginning the inquiry with a technical discussion of causality, it would be helpful to give an example of a causal account. Grasping the details and technical issues in the discussion will be easier with reference to this example. In choosing an example of a causal account, the choice is simple—

the most well-known part of Weber's work, the "Protestant ethic" thesis. Stated in its simplest and most commonly understood terms, Weber argues that: "Protestantism is the cause of the emergence of capitalism." This statement is simple in appearance only. The simplicity dissipates quickly as soon as one realizes that Weber means something quite specific by "Protestantism," something quite specific by "cause," and something quite specific by "capitalism." The way in which he frames the inquiry is also very interesting—What caused the emergence of a universally latent potential (i.e., "capitalism") to become actualized at a particular time and in a particular place in history (i.e., seventeenth-century northwestern Europe)? Weber notes:

> Insofar as our documents on economies have reached into the distant past "capitalism" and "capitalist" enterprises . . . have existed in *all* the world's civilizations. In other words, "capitalism" and "capitalist" enterprises have been found in China, India, Babylonia, Egypt, the ancient Mediterranean and medieval Europe. (PESC, 154)

Given the fact that capitalism is a universal phenomenon, Weber is intrigued by the fact that a unique species of this genus emerged at a particular time and place in history—even though it could have emerged in other environments. Weber's "Protestant ethic" thesis is not meant to provide a universal account for the emergence of capitalism—it seeks to identify the causal origin of the unique type of capitalism that emerged in the modern West. This becomes clear by looking at his precise definition of "capitalism" and the "'spirit' of capitalism":

> A fully unconstrained compulsion to acquire goods cannot be understood as synonymous with capitalism and even less as its "spirit." On the contrary, capitalism *can* be identical with the *taming* of this irrational motivation, or at least with its rational tempering. Nonetheless, capitalism is distinguished by the striving for profit, indeed, *profit* is pursued *again and again*, as is *profitability*. (PESC, 152)

Weber is aware of the fact that economic profit can be pursued in a variety of different ways and the "'spirit' of capitalism" demands a particular mode of conduct in the pursuit of economic profit. It requires the "taming" or "at least [the] rational tempering" of the irrational, compulsive desire to acquire economic wealth. Weber uses the term "rational capitalism" to refer to the type of capitalism that is underpinned by this "'spirit' of capitalism." He contrasts rational capitalism with "adventure capitalism." Weber describes the observable mode of conduct of adventure capitalists in these words:

> They have financed above all wars, piracy and all types of shipping and construction projects; as entrepreneurs in colonies they have served the interna-

> tional policy goals of nations. In addition, these adventure capitalists have acquired plantations and operated them using slaves or (directly or indirectly) forced labor; they have leased land and the rights to use honorific titles; they have financed both the leader of political parties standing for re-election and mercenaries for civil wars; and, finally, as "speculators" they have been involved in all sorts of money-raising opportunities. (PESC, 155)

While both rational and adventure capitalism pursue economic profit, the defining characteristic of adventure capitalism is that it does not recognize any internal or external constraints on the means that can be used to attain the end. In stark contrast, rational capitalism recognizes and accepts a number of external and (far more importantly) internal constraints on the means that can be used to purse profit. Weber notes that rational and adventure capitalism have existed side by side in a bewildering variety of mixtures in all human civilizations. For all of human history adventure capitalism has been much more prominent in the mixture than rational capitalism—with the unique exception of the early modern West.

Prior to the emergence of rational capitalism, traditional ethical norms were the primary means for putting limits on the means that could be used to pursue economic profit. Weber describes the dynamic between adventure capitalism and traditional norms in these words: "An absolute and willfully ruthless striving for profit often stood hard and fast alongside a strict adherence to age old traditions. Then, as these traditions began to disintegrate, a more or less broad-ranging expansion of an unrestrained quest for gain took place" (PESC, 21). Before their disintegration, traditional norms not only served as restraints on the adventure capitalist mode of conduct, they also blocked the development of rational capitalism. In other words, the emergence of rational capitalism was forestalled by factors other than the dominance of adventure capitalism. Weber identifies "economic traditionalism" as being the most significant of these factors: "In the sense of a certain norm-bound style of life that has crystallized in the guise of an 'ethic,' the spirit of capitalism has had to struggle primarily against a specific opponent: that type of experiencing, perceiving, and ordering of the world that we can denote *economic* traditionalism" (PESC, 21f.). Before giving a definition of "economic traditionalism," Weber gives an example. A farmer with a large field hires laborers to harvest the crop. He pays them one German mark for every acre harvested. During the course of a work day, the laborer harvests 2.5 acres and earns 2.5 marks. In the hopes of increasing productivity, the farmer increases their wage to 1.25 marks per acre—assuming that the incentive of higher wages will increase productivity and more acres will be harvested in one day. What the farmer finds is that after harvesting two acres, almost all workers stop working, collect their 2.5 marks and go home—the result being that fewer acres are harvested per day. Weber notes: "This example illustrates the type of behavior that should be called economic traditionalism. People do not wish 'by nature' to earn more and more money. Instead, they wish simply to live and to live as they have been accustomed and to earn as much as is required to do so" (PESC, 23). Weber finds this to be the "type of experiencing, perceiving, and ordering of the world" that is most commonly found among the largest segments of populations in most places in the world, at most times in history—prior to the emergence and triumph of modern capitalism.

In addition to economic traditionalism, Weber identifies particular religious teachings that were significant obstacles to the emergence of rational capitalism:

> The doctrine of *deo placer vix potest* [the merchant cannot be pleasing to God] was taken as genuine in medieval times . . . and was incorporated into canon law and applied to the activity of the businessmen. This was also evident in the view of St. Thomas Aquinas . . . who characterized the striving for profit as moral turpitude (including even unavoidable and thus ethically permitted profit). (PESC, 33)

In medieval Catholic piety, economic profit was spiritually suspect—to the degree of being spiritually harmful. One could not pursue economic profit and spiritual well-being at the same time. From the perspective of both economic traditionalism and religious piety, economic gain is something that one should pursue only as much as is needed to sustain life. The notion of pursuing economic gain in the sense of "profit" is utterly irrational from both of these perspectives. Adventure capitalism, economic traditionalism, and religious piety, singularly and collectively, created significant inner psychological barriers that had to be overcome in order for rational capitalism to fully develop. Weber's inquiry seeks to identify the causes that led to the overcoming of these inner barriers.

After describing how adventure capitalism had been the norm in most of the world for most of history, Weber notes: "However, in the *modern* era the West came to know an entirely different type of capitalism. Absent from all other regions of the globe, or existing only in preliminary developmental stages, this capitalism appeared side-by-side with adventure capitalism and took as its foundation the rational-capitalist organization of (legally) *free labor*" (PESC, 155). Weber's investigation sets out to discover the cause(s) for this unique historical occurrence—why did a universally latent potential (the triumph of rational capitalism over adventure capitalism) occur at a particular time in history and in a particular place? As a prelude to this inquiry, Weber sharpens the distinction between adventure capitalism and rational capitalism:

> Acquisition unrestrained by internally binding norms has existed in all periods of history; indeed, it has existed where its expression was not circumscribed. In relationships across tribes and among people fundamentally unknown to each other, as well as in warfare and piracy trade unbounded by norms has been the rule. A double standard prevailed in such situations: practices considered taboo "among brothers" were permitted with outsiders. (PESC, 21)

For Weber the pursuit of wealth "unrestrained by internally binding norms" is the defining characteristic of "adventure capitalism" or the "'spirit' of adventure capitalism." In stark contrast, the restrained pursuit of economic profit according to normatively binding norms, rules, and mores—free of double standards—describes the "'spirit' of rational capitalism." It needs to be emphasized that the restraints that are characteristic of the spirit of rational capitalism are the result of "internally binding norms" that the individual has voluntarily internalized. When Weber uses the term "capitalism" without any qualifier he is referring to rational capitalism and in this sense capitalism manifests itself in the form of "an *ethically*-oriented maxim for the organization of life. The expression *spirit of capitalism* will be used here in just this specific manner—naturally the spirit of modern capitalism" (PESC, 16). Weber identifies this "*ethically*-oriented maxim for the organization of life" as the source of a *Lebensführung* (mode of conduct) that distinguishes modern capitalism from other manifestations of capitalism:

> In light of the formulation of our theme, it must be evident that the Western European capitalism of the last few centuries constitutes our concern rather that the "capitalism" that appeared in China, India, Babylon, the ancient [Greco-Roman] world and the Middle Ages. As we will see, *just that peculiar ethic was missing in all these cases*. (PESC, 16)

Stated in more general terms, Weber's inquiry seeks to identify the particular cause(s) for the emergence of the *Lebensführung* that is uniquely characteristic of rational capitalism.

As is well known, Weber identifies a particular religious factor (the Protestant ethic) as the cause for the emergence of capitalism. In the terms used by Ola Agevall, the "'spirit' of capitalism" is the explanundum and the Protestant ethic is the explanan. Before detailing the evidence that Weber presents to support his position, a point of clarification on Weber's understanding of the relationship between effect and cause would be helpful. For Weber the spirit of capitalism (like any other unique individual in empirical reality) is the effect of an entire constellation of causes. Only the constellation as a collectivity can be properly labeled the cause of the effect. In the constellation of causes that produced the spirit of capitalism the Protestant ethic is only one element—with political, social, technological, geographical, and innumerable other elements also being a part of the constellation. The position that a particular event is the effect of an entire constellation of causes has to be reconciled with Weber identifying the Protestant ethic as the cause of the spirit of capitalism. The following is Agevall's formulation of Weber's thesis:"Certain denominations, within the Reformed Churches, were such that they, unintentionally, promoted a doctrine that was causally adequate for the rise of a *Lebensführung* identical to that of the 'spirit of capitalism', that is, a *Lebensführung* adequate to modern capitalism" (Agevall, 2009, 170). Stated in this way, Weber's Protestant ethic thesis "does not rule out the possibility that this *Lebensführung* could be produced in other ways" (Agevall, 2009, 170)—as a matter of fact, by the end of his investigation Weber himself explicitly identifies at least one other way (i.e., the "iron cage"). At the same time, this formulation is consistent with Weber's explicit statements that other constellations of causes "could bring about the same *Lebensführung* before modern capitalism had become rooted" (Agevall, 170) in the West. If this is the case—and for Weber it is indeed the case—then the question becomes even more pointed: In light of the fact that an entire constellation of causes produced the effect known as the "'spirit' of capitalism," on what grounds does Weber give the "Protestant ethic" a place of privilege in the constellation?

One possible answer is that the Protestant ethic produced a peculiar attitude toward worldly work. An important characteristic of the Protestant ethic was that it linked this-worldly striving with otherworldly salvation. On this point also, Weber points out that the Protestants were not the first ones to come up with a religious doctrine that linked salvation in the hereafter with hard and honest labor in this world. He notes: "The entire corpus of literature on asceticism, which is drawn from almost *all* religions, is permeated with the point of view that loyal work is highly pleasing to God, even if performed for low wages by people at a great disadvantage in life and without other opportunities. *Here* Protestant asceticism added nothing new as such" (PESC, 121). But this general characteristic that Protestantism shares with almost all religions takes on a unique characteristic when it is placed alongside the doctrine of worldly work as a vocation and the doctrine of predestination. In this new constellation the "view that loyal work is pleasing to God" is "dramatically deepened" by establishing a novel relationship between worldly work and otherworldly salvation: "[The asceticism of particular Protestant sects] created the norm on which its impact *exclusively depended*: the psychological *motivation* that arose out of the conception of work as a *calling* and as the means best suited (and in the end often as the sole means) for the devout to become certain of the state of salvation" (PESC, 121). While there are other examples of religious doctrines that link commitment to one's worldly work with otherworldly salvation (i.e., the notion of *dharma* in Hinduism), this relationship takes on a unique quality among the Reformed Protestant sects. For the Calvinist, Puritan, Methodist (and other sectarians) one does not make a vocational commitment to one's work because such a commitment will lead to salvation—rather one sees such a commitment as a sign (perhaps the only sign) that one has attained salvation in the world to come. Weber notes: "*Restless work in a vocational calling* was recommended as the best possible means to *acquire* the self-confidence that one belonged among the elect. Work and work alone, banishes doubt and gives certainty of one's status among the saved" (PESC, 66). And a little bit later: "One's state of grace is testified to through *conscientiousness* with which the believer, with care and methodicalness, pursues his calling. Rational work in a calling is demanded by God rather than work as such" (PESC, 108).

In addition to establishing a unique understanding of the relationship between worldly labor and otherworldly salvation, the Protestant sects came to have a unique view of the relationship between the pursuit of worldly wealth and otherworldly salvation. The Protestants "viewed the acquisition of wealth, when it was the *fruit* of work in a vocational calling, as God's blessing" (PESC, 116). Particular sects went further and reasoned that if wealth was indeed a sign of God's blessing then one should (keeping all the ethical strictures in mind) pursue it without any inner (or psychological) inhibitions. But at the same time one must keep in mind that worldly wealth (along with any other cultural good) was not to be enjoyed and indulged—this would be grossly displeasing to God. Weber describes the inner tension that these teachings created in the mind of the believer: "Ascetic Protestantism shattered the bonds restricting all striving for gain—not only by legalizing profit but also by perceiving it as desired by God. . . . The struggle against the desires of the flesh and the attachment to external goods was not, as the Puritans explicitly attest . . . a struggle against acquisition; rather it challenged the irrational use of possessions" (PESC, 115). At the same time that "the bonds restricting all striving for gain" are being shattered, the severest of restrictions are being placed on the use of these gains: "Persons are only administrators of the cultural performances that the grace of God has offered. Hence, every dime expended for them must be justified, just as in the example of the servant in the Bible. It remains doubtful at least whether any part of this money should be spent for a purpose that serves one's own pleasure rather than the glory of God" (PESC, 115).

Removing all inner inhibitions to the pursuit of profit and placing the strictest of conditions on the use of wealth conferred unique meaning and significance (*Sinn*) on economic activity. A vocational commitment to one's work and the accumulation of wealth serve as a sign (perhaps the only sign) that one is among those whom God has predestined to be among the elect. Weber sees this religious valuation of work and wealth playing a restraining role and a motivating role. In the former case it tamed the impulsiveness of adventure capitalism. In the latter case it shattered the hold of economic traditionalism and medieval religious piety. The result of restraining adventure capitalism and breaking the hold of economic and religious traditionalism was that an identifiable social class (various Protestant sects) adopted a particular mode of conduct (i.e., *Lebensführung*). It is worth quoting Weber at length as he describes the type of "capitalist" that Protestant asceticism produced:

> A specifically *middle-class vocational ethos (Berufsethos)* arose. Now the middle-class employer became conscious of himself as standing in the full grace of God and as visibly blessed by Him. If he stayed within the bounds of formal correctness, if his moral conduct remained blameless and if the use he made of his wealth was not offensive, this person was now allowed to follow his interest in economic gain and indeed *should* do so. (PESC, 120)

Not only did the Protestant *Berufsethos* give rise to a particular type of owner of a capitalist enterprise, it also gave rise to a particular type of worker: "Moreover, the power of religious asceticism made available to the businessperson dispassionate and conscientious workers. Unusually capable of working, these employees attached themselves to their work, for they understood it as bestowing a purpose in life that was desired by God" (PESC, 120). At the conclusion of his discussion linking Protestant religious beliefs with capitalist economic enterprise, Weber notes: "It is obvious how powerfully the exclusive *striving* for the kingdom of God . . . must have prompted the productivity of work in the capitalist sense of the word" (PESC, 121). Judging from their outward behavior, it is difficult to determine which of the two is more committed to the pursuit of profit, the adventure capitalist or the Protestant capitalist and worker. But the particular "ethic" that is adopted in the pursuit of profit sets the adventure and rational capitalist on the two different ends of the spectrum. And for Weber, among the entire constellation of factors that are collectively the cause of the spirit of capitalism, this particular factor (the "ethic" of Protestantism) holds a special place. The values that are inherent in the observable mode of conduct are in their turn underpinned by a set of personal beliefs. The investigator must properly identify these subjective beliefs on the basis of observable conduct and values if he is to adequately interpret the actions of the actor. (In chapters 3 and 4 we will see how crucial beliefs and values are, not only in the construction of socio-cultural reality but also in the scientific analysis of this reality. For the time being, suffice it to say that there is no such thing as socio-cultural reality or its scientific investigation in the absence of beliefs and values.)

It was noted earlier that the Protestant ethic is only one element in an entire constellation that Weber identifies as the cause for the emergence of capitalism. The question naturally emerges as to why Weber accords a place of privilege to the Protestant ethic as the most significant causal factor. In order to understand this we have to bring in the perspective of the investigator. From the point of view of the question that initiated the inquiry, the Protestant ethic is the single most important causal factor from the entire constellation of causes. In other words the relationship between the explanundum and the explanan is established by the investigative concern of the investigator. Weber's investigative concerns in the PESC are not made explicit in the beginning of the work. They are actually not made explicit until he writes an "Introduction" in 1920 to a collection of essays in the sociology of religion, of which the PESC is a part. The opening paragraph of the "Introduction" identifies Weber's key concerns:

> Any child of modern European culture will, unavoidably and justifiably, address universal-historical themes with a particular question in mind: What combination of circumstances led in the West and only in the West to the appearance of a variety of cultural phenomena that stand—at least as we like to imagine—in a historical line of development with universal significance and empirical validity? (PESC, 149)

One of the reasons that Weber undertakes a study to identify the origins of modern capitalism is to understand the "combination of circumstances" that produced the unique characteristics of modern European culture. This line of inquiry is all the more intriguing for Weber because practically every unique characteristic of European culture (with modern capitalism being one such characteristic) is found in its underdeveloped form in cultures across the globe. Between the lines (but just barely) Weber is saying that there is a far more pressing reason to take up this line of inquiry than intellectual curiosity. Understanding the universal significance and explaining the empirical validity of these unique characteristics is not possible without a scientific account of their causal origins and historical development.

In Weber's day modern capitalism was among the most widely recognized unique characteristics of Western culture. His investigative concern is to identify those causal factors in Western history that facilitated the emergence of modern capitalism in the West and hindered its emergence in other cultures. In this context, the Protestant ethic becomes a crucial causal factor in the emergence of capitalism in the West. As we will see in the following chapter, the other causal factors for the emergence of capitalism were present in other cultures centuries earlier than northwestern Europe but the growth of capitalism remained stunted—hence the importance of the Protestant ethic as a causal factor for Weber.

Weber's investigation also shows that with the passage of time the cause (Protestant ethic) of the effect (capitalism) has been profoundly affected by the effect and has changed dramatically. This in its turn has profound consequences for the culture at large. In summary form, at one point in history (when the Protestant ethic was vibrant) capitalist activity was pursued as a matter of an "inner calling" (*Beruf*). After originally germinating, taking root, and growing into a strong, healthy tree in the soil of the Protestant ethic, the spirit of capitalism has produced fruit that has profoundly changed the character of the soil in which the tree originally took root. The most important change in this regard, from Weber's perspective, is the fact that a cultural activity that was once engaged in because of voluntary inner conviction, must now be carried out due to overbearing external compulsion. Weber captures this shift from inner conviction to external compulsion, in these (oft-quoted) words:

> The puritan *wanted* to be a person with a vocational calling; today we *are forced* to be. For to the extent that asceticism moved out of the monastic cell, was transferred to the life of work in a vocational calling, and then commenced to rule over this-worldly morality, it helped to construct the powerful cosmos of the modern economic order. Tied to the technical and economic conditions at the foundation of mechanical and machine production, this cosmos today determines the style of life of all individuals born into it, *not* only those directly engaged in earning a living. This pulsating mechanism does so with overwhelming force. Perhaps it will continue to do so until the last ton of fossil fuel has burnt to ashes. According to Baxter, the concern for material goods should lie upon the shoulder of his saints like "a lightweight coat that could be thrown off at any time." Yet fate has allowed [an iron cage][1] to be forged from this coat. (PESC, 123)

The change in the external conditions in which the spirit of capitalism manifests itself has brought with it a change in the ethic on which the spirit is based. This is another way of saying that with the radical change from inner conviction to external compulsion, the subjective *Sinn* (meaning) that the actors invest in the observable activity of the pursuit of profit has also changed radically.

In line with Weber's observation of the "particular question" that "any child of modern European culture" will ask when looking at "universal-historical themes," we can safely say that Weber wants to clearly delineate the causes and the consequences in the profound shift in *Sinn*. This diagnostic exercise is an essential prerequisite for any attempt to remedy the cultural malaise that the shift may have engendered. We know from his other writings that as a "child of modern European culture" Weber is deeply troubled by certain developments. Rational capitalism along with modern science, liberal democracy, and constitutional law are among the ideals that Weber holds dear. These ideals, which have shaped modern Western culture, are the product of a unique constellation of historical factors. One of the most important parts of this constellation is the commitment to certain values on the part of the cultural beings living in the historical constellation. With the passage of time it has become increasingly difficult to affirm these value commitments. This has significantly changed the character of the constellation in which the ideals first emerged so much so that the future well-being of these ideals is in doubt. A scientifically objective analysis of the "line of historical development" is needed to offer a scientifically valid (or sober) diagnosis and workable remedy for the present-day developments that are undermining the values that Weber finds worth affirming. In sum, the investigative concerns shaping Weber's inquiry into the origins of rational capitalism are twofold: (1) understanding the historical origins of the cultural values that the investigator considers to be significant and valid; and (2) understanding the present-day developments that are undermining the health and well-being of the cherished values.

Among an entire constellation of causal factors responsible for the emergence of the "'spirit' of capitalism" Weber identifies commitment to a set of values that he calls the "Protestant ethic" as an especially significant factor. In summary form his causal account goes something like this: At a particular time in history, at a particular place certain cultural beings made a decision to affirm the "Protestant ethic," this decision was the cause of the rise of the "'spirit' of capitalism." Among his contemporaries and predecessors a number of other theories had been offered to explain the same phenomenon. In broad terms, one explanation was that the rise of capitalism was caused by the "laws of economics" (similar to the claim that the "laws of gravity" or the "laws of nature" cause objects to fall to the ground). Another explanation was that something inherent in "human nature" or the "nature" of Western culture caused the emergence of capitalism (similar to the claim that the "true nature" or "essence" of something is the cause of its observed action or behavior). Weber rejected both explanations on logical and methodological grounds. Before engaging in second-order reflections on Weber's causal account, it will be worthwhile to look at his critique of these positions.

ROSCHER, LAWS, AND CAUSALITY

Wilhelm Rosher belonged to the "Historical School" of economics. His ideas are of relevance to the present discussion because his position on the topic of causality places him in the "positivist" school—those who attributed causal agency to the "laws of nature" (or the "laws of history" as is presently the case). The Historical School contrasted itself with the Classical School of economics because the latter was content with discovering "in the multiplicity of events the law-like and uniform rule of simple forces" (RK, 58). The Historical School for its part took on a different task in light of the fact that "the multiplicity of events" itself undergoes a process of change and evolution. Consequently for Roscher the goal of economics as a science "is to establish evolutionary laws of *historical* change" (RK, 60). By way of analogy, the Classical School followed the Newtonian model and sought to identify the universal laws that allegedly govern economic institutions, procedures, and behavior. In contrast the Historical School followed the Darwinian model and sought to identify those immutable laws of history/economics that allegedly govern the evolution of society. The concept of "Volk" was central to Roscher's theory of historical evolution. Weber notes that Roscher "conceives the 'Volk' in terms of an analogy he draws between the development of the life of an individual human being and the development of culture: the 'Volk' is the *individual*, which experiences the gradual developments of the state, law, and economy" (RK, 61). After noting that Roscher "does not subject the concept of 'Volk' to any further analysis" but at the same time "does not conceive it as an abstract generic concept, relatively empty of content" (RK, 62), Weber draws the following conclusion from Roscher's analysis: "The purely rationalistic concept of the 'Volk' as the coincidental collectivity of politically associated citizens does not satisfy him. In place of the generic concept, which is a product of abstraction, he employs the following conception of the 'Volk': It is an *intuitable* totality, the cultural bearer of a *meaningful* total essence" (RK, 62). While Roscher fails to subject his construction and use of the concept of Volk to logical scrutiny, Weber takes up this task with purposeful focus. Weber makes explicit the process by which the concept of Volk emerges in Roscher's work:

> Roscher believes that he can employ the unanalyzed concept of an intuitable multiplicity of "Volk" in the same way that a biologist works with the unanalyzed concept of the perceptual manifold "elephant." The various "Volk," he thinks, are in reality just as different from one another as individual human beings. The latter difference does not prevent anatomists and physiologists from ignoring them in their observations. (RK, 63)

After describing the process by which Roscher arrived at the concept of "Volk" (following the methods of the biologists, anatomists, and physiologists) Weber goes on to describe the role that this concept will play in helping Roscher establish the "evolutionary laws of historical change." The parallel with the method of the natural sciences that ignore the unique characteristics of the phenomenon being studied is clear:

> The individual peculiarities of nations do not prevent the historical theorist from treating nations as members of a class and comparing their development in order to establish *correlations*. Roscher thinks that with the progressive completeness of observation, these correlations will eventually be elevated to the logical status of "natural laws" which hold for all members of the class "Volk." (RK, 63)

At this point Weber draws our attention to the value and limitations of the "correlations" and "natural laws" that are discovered in this manner: "In individual cases, a complex of regularities discovered in this fashion may have extraordinary *heuristic* value. However, it should be obvious that their discovery cannot be conceived as the ultimate *goal* of any science: neither a 'nomological' science, nor an 'historical' science, neither a 'natural' science nor a 'sociocultural science'" (RK, 63). The primary reason why the discovery of correlations, recurring patterns (often called "nomological" regularities in Weber's milieu) or laws, cannot be "the ultimate *goal* of any science" is because these "generalizations would not have *causal* status" (RK, 63). But Roscher does attribute causal status to these conceptual generalizations (along with other thinkers in the Historical School of economics and in agreement with the Classical School that they are critiquing on other points): "Roscher has no doubt, in principle, that the relations between economic phenomena can be and should be conceived exclusively as a system of *laws*. 'Causality' and 'nomological regularity' are identical from his point of view. The first exists only in the form of the second" (RK, 60). Weber's analysis and critique of Roscher's ideas is only one example of a theme that is repeated often and regularly in Weber's methodological writings—the value and limitation of "laws" in scientific inquiry (irrespective of whether they are laws of nature, history, mind, etc.). Weber identifies the source of the widespread belief among the hard scientists (and their imitators in the soft sciences) that "laws of nature" (or some equivalent term) can be invested with causal agency in these words: "It is a consequence of the bioanthropological aspects of the various influences which the atrophied remains of the great Hegelian ideas have exercised upon the philosophy of history, language, and culture" (RK, 207). For Weber such a theory of "causality" in scientific inquiry "can only produce results that are essentially negative and even destructive" (RK, 207). In sum, Weber's position is that correlations, nomological regularities, laws (in short, generalizations of any type) "have extraordinary *heuristic* value" in scientific inquiry but "have no *causal* status."

Toby Huff notes that the details of Weber's critique of Roscher's position are of more than historical interest. The logic underpinning Weber's analysis and critique of Roscher's position contains within it the critique of some varieties of inductivism and deductivism that became popular after Weber's death (Huff, 38f.). Weber uses the opportunity afforded by his engagement with Roscher to offer a more general comment on the deductivist approach to science. Referring to Roscher's claim that "progressive completeness of observation" will eventually yield "a system of formulae of absolute general validity" that would be sufficient to offer causal accounts of empirical phenomena, Weber notes:

> This is not only empirically impossible. Because of the logical nature of "nomological" knowledge, it is also logically impossible. The formation of "laws"—relational concepts of general validity—is identical with the progres-

> sive depletion of conceptual content through abstraction. The postulate of the "deduction" of the content of reality from general concepts . . . is logically absurd, even as an ideal the realization of which lies in the indefinite future. (RK, 218, n.23)

In Weber's critique of Roscher's deductivism of the 1840s, Huff sees valid grounds for the critique of two varieties of deductivism that became popular in the 1950s—the hypothetico-deductivism of Popper and the covering law model deductivism of Hempel (Huff, 38).

Furthermore, Huff sees the following passage by Weber as a pointed and valid critique of the inductivist claim that "the 'laws' in science are but summary statements of observations" (Huff, 38). After allowing for the theoretical possibility that history and other soft sciences "could establish an enormous number of 'empirical' historical generalizations" (i.e., laws), Weber goes on to note: "Nevertheless, these generalizations would have no *causal* status. Correlations of this sort could only constitute the subject matter of a scientific investigation, an investigation which would begin only after these correlations had been established" (RK, 63f.). Once the task of establishing the correlations has been completed it is only at that point that the crucial question in the course of scientific inquiry can be asked: "At this point—and this is most important—the investigation would have to produce a decision concerning the following issue: What sort of knowledge should be the aim of the investigation?" (RK, 64). For Huff, this part of Weber's analysis should be considered with utmost care and diligence because: "Weber puts us on notice that whatever 'empirical generalizations' might emerge out of research *must themselves* be explained—not simply organized into shopping lists—according to criteria that are by no means self-evidently reflected in the phenomena themselves" (Huff, 38). In this part of the discussion Weber has stood the fundamental claim of both the inductivists and the deductivists on its head. For both of these positions, the ultimate goal of all scientific inquiry is the discovery of "laws" that (allegedly) explain, give reasons for, or identify the causes of empirical phenomena. We can summarize Huff's evaluation of the deductivist and inductivist philosophies of science in light of Weber's pre-WWI insights by noting that for Weber all such theories eventually "fall victim either to Hegelian emanatism or to some variety of modern anthropological occultism" (RK, 194).

The fact that deductivism and inductivism (in all of their varieties and forms) "fall victim" to some sort of occultism or other is by a logical analysis of the ultimate "end" that they aim for. If discovery of the "laws of nature" (whether by inductive means or deductive means) is the ultimate goal of scientific inquiry then science is caught in a debilitating conundrum—the "laws" that allegedly provide the explanations, reasons, and causes of empirical reality remain incapable of explanation, devoid of reason, and without any cause(s). In short, the laws are rendered scientifically irrational. For Weber the fact that the laws themselves are rendered irrational by the inductivist and deductivist approaches is evidence enough that the discovery of laws cannot be the end of scientific inquiry. For Weber this "end" is actually an intermediate stage in scientific inquiry because the ultimate goal is to discover the reason and causes of these laws (i.e., to make these laws rational and explicable). At this point Huff cites a brief passage from the work of Charles Peirce to show the similarity between Weber and Peirce on this point. It is worth quoting the complete passage by Peirce to better understand this point. Speaking about regularities, laws, and reasons, Peirce notes:

> To suppose universal laws of nature capable of being apprehended by the mind and yet having no reason for their special forms, but standing inexplicable and irrational, is hardly a justifiable position. Uniformities are precisely the sort of facts that need to be accounted for. That a pitched coin should sometimes turn up heads and sometimes tails calls for no particular explanation; but if it shows heads every time, we wish to know how this result has been brought about. Law is *par excellence* the thing that wants a reason. (6.12)

Taken together, Weber and Peirce offer an understanding of "scientific" inquiry that significantly modifies the naïve understanding of the relationship between "laws" and empirical reality in the dominant inductivist and deductivist philosophies of science. This naïve understanding attributes the reason (or cause) of empirical phenomena to "universal laws of nature capable of being apprehended by the mind." But at the same time it becomes logically impossible to engage in a scientific inquiry to discover the cause(s) or reason(s) for the "special forms" that these laws have taken. In other words we cannot ask the question "Why are the 'laws of nature' as they are, when they could have been otherwise?" If this question cannot be asked about a particular phenomenon, then for Weber that phenomenon is put outside the scope of scientific inquiry. Putting an important (perhaps a unique) aspect of scientific inquiry called the "laws of nature" outside the scope of scientific inquiry is ultimately the unscientific result of a line of reasoning that begins with the occult, unscientific claim that the "laws of nature" have causal agency.

WUNDT, HUMAN CREATIVITY, AND CAUSALITY

While the positivism of the Historical School attributed causal agency to the "laws of nature" and made the discovery of "laws" the ultimate end of scientific inquiry, the intuitionism of the Verstehen School attributed causal agency to something unique in human nature that was not to be found anywhere in the world of nature. While they differed with the Verstehen School on other issues, Wilhelm Wundt and Karl Knies are examples of this intuitionist position, with the former identifying human creativity and the latter identifying human freedom as uniquely human characteristics. Both Wundt and Knies go on to articulate arguments that attribute causal agency to that which is unique to "human nature" and is not found anywhere in physical-natural phenomena. Weber notes: "We repeatedly find reference explicit or implicit to the unpredictability of personal conduct. This is alleged to be a consequence of 'freedom,' the definitive source of human dignity. . . . At the same time, a distinction is drawn between the 'creative' significance of the acting personality and the 'mechanical' causality of natural events" (RK, 97f.). Wundt claims that human thought and action (or the "human personality") are endowed with uniquely "creative" capacities that are absent from the inert things and mechanical, deterministic processes in the natural world. Wundt sees human thought and action as being the manifestation and the grounds of "creativity"—with "creativity" being characterized by novelty and spontaneity. For Wundt, there is a direct correlation between this "creative capacity" and causal agency—causal agency can be only found where there is "creative capacity." Consequently, Wundt claims that human beings alone have causal agency. In contrast to the human world, the natural world is characterized by mechanistic repetition and determinism where "causal agency" is completely absent. Wundt distinguishes the soft sciences from the hard sciences on the basis of the absence or presence of this creative capacity–causal agency. Weber demonstrates that Wundt's understanding of "creative" is only partially right—and the failure to recognize and redress its inadequacy is detrimental to scientific inquiry.

An important corollary of Wundt's argument is that since creativity, novelty, and spontaneity are found exclusively in the socio-cultural domain, it naturally follows that *Sinn* belongs exclusively to this domain also. One could infer from Weber's critique of Wundt's "creative synthesis" claim that Weber would reject the latter claim also—but one does not have to go through the trouble of inferring. Along with demonstrating the shortcomings in Wundt's human creativity and causal agency claims, Weber demonstrates (using Wundt's own logic) that *Sinn* is as much a part of physical-natural phenomena as it is of socio-cultural phenomena.

Weber notes that Wundt uses the term "creative synthesis" in a number of different ways, and he subjects each of the different uses to critical analysis. Irrespective of the sense in which Wundt uses the term, Weber's conclusion is the same—there is nothing specifically "creative" about human thought or action in contrast to the "mechanical" processes in the natural world. One of Wundt's uses of "creative" is in the sense of "producing something new." Weber demonstrates that we find "creative synthesis" at work in the physical-natural world no less than in the socio-cultural world: "From a logical point of view, the physical and chemical processes which produce a seam of coal or a diamond constitute a 'creative synthesis' in the same sense as the chain of motives which links the intuitions of a prophet to the formation of a new religion" (RK, 102). If we view the developmental processes leading up to the emergence of a "seam of coal" or "a new religion" strictly as a process of causes and effects, then any particular seam of coal and any particular religion are equally products of a "creative synthesis." At the end of both processes we find the emergence of something novel that is not present at the beginning or in the middle of the course of development of the phenomenon under study. When we look at the constituent elements in the causal process it is obvious that the "human" element is completely missing in the case of coal, diamonds, oil, trees, bees, or any other natural phenomenon.

Since "creativity" is as much a part of physical-natural world as it is of socio-cultural world, one has to be exceedingly careful when offering causal accounts of events. Weber notes that the human agent is only one of many (actually innumerable) causal factors involved in any event that takes place in the socio-cultural world. To give the human agent a privileged place in the causal account says more about the investigative interests of the investigator than anything else. Weber notes: "Suppose we consider the special properties and the concrete action of a concrete 'historical' personality 'objectively'—that is, independent of our specific interests. In this case, there is no sense in which such a personality is a 'more creative' *causal* factor in history than any 'impersonal' causal factor—geographical or social conditions or specific natural processes—*can* be" (RK, 101). As Kalberg has demonstrated in his skillful reconstruction of Weber's account of the rise of monotheism in the ancient Near East, climate (rainfall or lack thereof) and technology (building and maintaining vast irrigation systems) played as significant a causal role as the personality of charismatic Prophets (Kalberg, 1994b).

In addition to using the term "creative" in the sense of producing or giving rise to something new, Wundt uses the term in the sense of adding a novel quality to something that is already present. Wundt argues that human thought is "creative" in a way that natural phenomena are not. In simple terms, for Wundt, human thoughts are always more than the sum of their parts, whereas phenomena of nature are never more than the sum of their parts. To support his claim Wundt offers the example of a crystal and a biological organism. For him both of these things can be described purely in terms of their constituent elements. Weber summarizes Wundt's position, using Wundt's own words as follows: "From the point of view of the natural scientist, a crystal can be 'nothing more' than 'the sum of its molecules and their external reciprocal effects.' From this point of view, the same holds for an organism. The natural scientist may not yet be able to 'causally deduce' 'the whole' from its elements. Nevertheless, 'the organism is a product of these elements. They contain a complete prototype of the organism itself'" (RK, 106). For Wundt, the crystal and the organism do not contain any "new properties" that are not already present in their constituent elements. Weber challenges this claim by asking us to consider the properties and characteristics of the solar system: "Consider the purely quantitative relations of the solar system, its 'elements' being the independently identifiable individual planets or the mechanical forces which possibly generated the solar system from a hypothetical primeval gas" (RK, 105). A careful consideration of these elements reveals that the solar system contains numerous "new properties" that are nowhere to be found in its constituent elements. This "in spite of the fact that the solar system can be described in terms of a chain of concrete, purely physical processes, each of which can be reduced to an expression of causal equivalences" (RK, 105).

Weber sharpens his critique by drawing attention to the fact that when we focus our attention on qualitative properties, the shortcoming in Wundt's position is even more obvious. Weber agrees with Wundt that "mental structures" do indeed contain qualitative properties that are not present in their constitutive elements. But, for Weber, it is equally obvious: "That this also holds in the same sense and to the same degree for all natural processes conceived as qualitative changes. For example, water, from the point of view of its qualitative peculiarities, has properties which are simply not 'contained' in its constitutive elements" (RK, 105). It is obvious that water contains entirely novel qualitative properties that are not (and can never be) found in hydrogen, oxygen, and any catalyst used to bring the two together. It is worth emphasizing that this is true for "all natural processes" not just water. Weber goes on to note that besides quantitative and qualitative change, "whenever the question of value relevance enters the picture, there is *no* natural process which fails to contain specifically 'new' properties not possessed by its 'elements'" (RK, 105). This point is even more obvious than the fact that water contains qualities that have not been found, are not found, and will never be found in oxygen, hydrogen, and any catalyst that bonds the two. When a particular value is attached to a particular phenomenon, for Weber a peculiarly "new" property emerges—*Sinn*. This property is something that is not inherent in either the object or the value placed upon the object but is the result of a particular object being brought into relationship with a particular value. The example of water can be used to illustrate this point also. In terms of its origin, water is a completely "natural" object that comes into existence as a result of completely "objective, natural" processes. Water has been valued variously as medicine, sacred purifying agent, profane cleansing agent, sacred symbol of divine, profane symbol of political authority, cooking ingredient, instrument for torture, etc., by human beings in different cultural settings. Weber would say that the *Sinn* (in the sense of "significance") of water has varied in history. It is also a fact that water is completely unknown to human beings independent of some value that they attach to it.

Using Wundt's own words, Weber summarizes his position on how and why a mental construct is more than the sum of its parts:

> An "idea," [Wundt] claims, is "never simply the sum of the perceptions into which it can be analyzed." Further, he describes "intellectual processes"—for example, a judgment or a conclusion—as structures which can never be conceived "as mere aggregates of single perceptions and ideas." For, he adds, "that which is required in order to give these processes *meaning* is produced . . . *out* of these components. However it is not contained *in* them." (RK, 107)

Weber notes that this is indeed the case—the meaning (*Sinn*) that humans invest in their ideas, judgments, and conclusions is never "contained *in*" the processes that generate the various mental structures. *Sinn* is something that is "creatively" added to these processes—when particular thoughts are brought into relation with particular values. But Weber goes on to note that the relationship between human beings and their ideas is no different from the relationship of human beings and the natural world. After summarizing Wundt's position on how human beings attach meaning to their ideas, judgments, and conclusions, Weber states: "This is true enough. But is the formation of the 'products of nature' any different? Consider the 'meaning' of the diamond or the stalk of grain for certain human 'feelings of value.' Is this 'meaning' 'implied' to a higher degree or in another sense in the physicochemical conditions of their developments than is the case . . . as regards the 'elements' out of which ideas and judgments are formed?" (RK, 107). Weber's point is clear and consistent enough. If the "formation" of ideas and judgments is the result of meaning being attached to them, then things, events, and processes in the world of nature are "formed" in the same way. In both cases meaning (*Sinn*) is something that is added to the phenomenon independent of its constituent elements and developmental processes. He emphasizes this point a bit later in the discussion by using different examples: "The origin of the solar system out of some primeval gas, or—if the legitimate application of this concept depends upon the suddenness of the event to be explained—the incursions of the Dollart, *can* fall under the concept of the "creative" just as easily as the Sistine Chapel Madonna or the original conception of Kant's *Critique of Pure Reason*" (RK, 119f.). This observation by Weber establishes the fact that even if "creativity" is understood as "spontaneity" rather than "newness" or the generation of *Sinn,* then there is no difference between socio-cultural phenomena and physical-natural phenomena—the solar system is as much an outcome of a "creative" process as *Critique of Pure Reason*.

Since the meaning that human beings confer on empirical phenomena cannot be deduced from the constituent elements or processes of causation, the question naturally emerges regarding the origin of this "creative" element. For Weber the term "creative synthesis" is meaningful only in the context of ascribing meaning to "facts" in the world by bringing facts (whether they are socio-cultural or physical-natural) into relation with "values." Weber notes:

> The *meaning* we ascribe to phenomena—that is, the relations we establish between these phenomena and "values"—is a logically incongruous and heterogeneous factor which cannot be "deduced" from the "constitutive elements" of the event in question. The process whereby "we" associate "mental" events [or natural events] with values—regardless of whether they are diffuse "feelings of value" or rational "value judgments"—constitutes the "creative synthesis." (RK, 108)

And a little bit later: "The concept of the 'creative' has a function only when we begin to relate individual components of series of changes—which are 'intrinsically' of no significance at all—to *values*" (RK, 119). It is obvious that *Sinn* becomes a reality when an interpreting agent brings two things that are "intrinsically" of no significance, (i.e., a particular fact and a particular value) into relation with each other. Going back to Wundt's position, Weber notes: "In the work of Wundt, astonishingly enough, these relations are reversed. In his view, the principle of 'creative synthesis' is 'objectively' grounded in the specific properties of mental causality. This principle finds its 'characteristic *expression*' in definitions and judgments of value" (RK, 108). By doing this Wundt attempts to make meaning (*Sinn*) exclusively a characteristic of socio-cultural phenomena and render physical-natural phenomena meaningless.

In the *Critique of Stammler* (especially pages 107–115), Weber discusses the relationship between nature, social life, and *Sinn* in detail. He undertakes this discussion to demonstrate a simple but most important fact: no matter how one defines "nature" (or "social life"), a consistent application of the definition reveals that *Sinn* is as much a part of physical-natural phenomena as it is of socio-cultural phenomena and it is generated by the same reasoning processes. It is worth quoting Weber at length on this point:

> Suppose that for conceptual purposes we distinguish the "meaning" which we find expressed in an object or process from all the other components of the object or process which remain after this "meaning" is abstracted from it. And suppose that we define the sort of inquiry that is *exclusively* concerned with this last set of components as "naturalistic." In this sense of "nature," nature is the domain of the "meaningless." Or, more precisely, an item *becomes* a part of "nature" *if* we cannot raise the question: What is its "meaning"? Therefore it is self-evident that the antithesis of "nature" as the "meaningless" is not social life, but rather the "meaningful": that is, the "meaning" ascribed to a process or object, the "meaning" which can "be found in it." This includes, at one extreme, the metaphysical "meaning" of the cosmos as seen from the perspec-

> tive of a certain system of religious dogmatics. At another extreme, it includes the "meaning" which the barking of Robinson Crusoe's dog "has" when a wolf approaches. (CS, 110f.)

Whether one is speaking of the "meaning" of the universe or the "meaning" of the barking of a dog, in both cases "meaning" is generated by bringing an empirical fact (universe or barking) into relationship with a value by an interpreting agent. Weber combines logical analysis and commonsense examples to offer the following conclusion: "Therefore, we have shown that there is no sense in which the [quality] of being 'meaningful'—the property of 'meaning' or 'signifying' something—is a feature peculiar to 'social' life, its definitive property" (CS, 111). Consequently, whether we are talking about "objects" in the socio-cultural domain or in the physical-natural domain, we find that any object that becomes the focus of scientific inquiry is composed of three elements:

1. Fact (objectively observed in empirical reality)
2. Value (subjectively held by the actor)
3. Meaning (brings fact and value into relationship)

In sum, Weber's analysis and critique of Wundt's position demonstrates that causal agency cannot be attributed exclusively to human creativity/personality—"impersonal" physical-natural phenomena are also causal factors in concrete, individual events. Similarly, "there is no sense in which the [quality] of being meaningful" is an exclusive characteristic of socio-cultural reality. From a strictly logical point of view, to the degree that the physical-natural world is an object of scientific inquiry, it is no less suffused with meaning and significance than the socio-cultural world.[2]

KNIES, HUMAN FREEDOM, AND CAUSALITY

The ideas of Karl Knies are a variant of Wundt's position. The similarity between the two is that they attribute causal agency to something that is unique to "human nature." The difference is that whereas this something unique is human creativity for Wundt and it is human freedom for Knies. Even though the difference between the two positions appears insignificant at first glance, a closer look reveals that Knies's position requires separate treatment because it raises a different set of logical and methodological issues. Wundt's human creativity position required Weber to demonstrate that novelty and meaning are not exclusive characteristics of the socio-cultural domain. Knies's human freedom position requires Weber to demonstrate that incalculability and irrationality are not unique characteristics of the socio-cultural domain. Knies puts forward the claim that causal interpretation of events in the natural world differs from causal interpretation of events in the human world because human action is uniquely characterized by "incalculability" which is "the characteristic mark of human 'freedom of the will'" (RK, 120). Consequently, Knies employs the notion of the incalculability of human action "in order to provide a basis for the peculiar status of 'sociocultural sciences,' a status which is a consequence of the following: *because of this incalculability*, these disciplines are concerned with an entity that has an intrinsic dignity" (RK, 120). Weber notes that from a purely logical perspective, "'experienced' concrete reality contains no trace at all of a species of 'incalculability' *peculiar to* human conduct" (RK, 120f.). Weber supports his position by drawing our attention to the fact that all human interaction and communication is based on the presupposition of the calculability and predictability of human action (and human motives):

> Every military order, every criminal law, in fact every remark that we make in conversation with others, "counts" on the fact that certain impressions will penetrate the "psyches" of those for whom they are intended. They do not depend upon the absolute unambiguity of these impressions in every respect and in every case. But they do depend upon a calculability which is sufficient *for the purposes* which the command, the law and the concrete utterance are intended to serve. (RK, 121)

For Weber a logical analysis of the "calculations" that are used to interpret "every military order, every criminal law, in fact every remark that we make" reveals that these "calculations" are no different from the calculations that we make when interpreting data derived from the study of physical-natural phenomena:

> From a logical point of view, these calculations are no different from the "statistical" computations of a bridge builder, the agricultural-chemical calculations of an agricultural economist and the physiological hunches of a stock breeder. And the last are "calculations" in the same sense as the economic estimates of a strike organizer or a labor negotiator. Each of these "calculations" is satisfactory if it produces the empirically attainable degree of "precision" which is necessary and sufficient, given the specific purpose and the condition of the source material. There is no logical difference between these cases and "natural processes." (RK, 120f.)

In sum there is no logical difference between "calculations" involved in interpreting three types of phenomena that cut across the divide of sociocultural and physical-natural domains: (a) human conduct and utterances, (b) "statistical computations" involved in interpreting the natural world, and (c) "economic estimates" involved in interpreting the trends in the marketplace.

Weber's critique of Knies goes further than demonstrating that there is no peculiar type of unpredictability that is unique to human action. When it comes to providing a causal account of particular individual events in the future, socio-cultural phenomena are more predictable than physical-natural phenomena:

> The "calculability" of "natural processes" in the domain of "weather forecasting," for example, is far from being as "certain" as the "calculation" of the conduct of a person with whom we are acquainted. Regardless of the completeness of our nomological knowledge, [calculation in weather forecasting] is still not susceptible to the same degree of certainty. The same holds whenever the concrete uniqueness of a *future* "natural process"—as opposed to specific, abstracted relations—is at issue. (RK, 120)

Our "calculation" of what one of our close acquaintances will be doing at a particular hour, on a particular day, during a particular month in the future is likely to be far more accurate than our prediction of what the weather will be like at a particular hour, on a particular day, during a particular month in the future. This example evidences that when it comes to predicting unique, individual future events there is a significant degree of difference between calculability of socio-cultural phenomena and natural-physical phenomena—with the former being far more calculable. Weber goes on to demonstrate that the difference between the two reaches a point where the difference is no longer a difference in degree but becomes a difference in kind—where one can attain a kind of predictability of socio-cultural phenomena that simply cannot be had of physical-natural phenomena.

Weber asks us to consider an example from the physical-natural world in which a boulder falls off a cliff during a rainstorm. Given ideal conditions of observation and taking "explainable" to mean "verification after the fact" for this event: "The following are all causally 'explainable'—albeit rather imprecisely—in terms of established laws of mechanics: the occurrence and the general direction of the fall; the occurrence and perhaps the general angle of the splintering; also the general direction of one fissure or another" (RK, 122). While there are a number of questions that a scientific analysis of the falling boulder can answer, there is a vast category of questions that cannot be answered. Weber explains:

> But consider the following questions. Into how many fragments was the boulder broken? How are the fragments shaped? And into what groups did the scattered fragments fall? As regards these and a veritable infinity of similar "aspects" of the phenomenon, even though they can be described in terms of purely quantitative relations, our criteria for causal explanation—*assuming* that for some reason knowledge of these aspects of the event was important—would be satisfied by the following conclusion: the established facts are not "inexplicable." [In other words] they entail no consequences which are *inconsistent with* our "nomological" knowledge. (RK, 122)

For Weber there is a "veritable infinity" of aspects of this particular event that can be explained only by the statement that "they entail no consequences which are *inconsistent with* our 'nomological' knowledge."

In addition to the splintering boulder example, he offers the example of "a concrete [i.e., individual, particular] throw of the dice" (RK, 126). What type of scientific knowledge can be attained about the results of an individual particular throw of the dice which turns up a 6? Weber explains: "The fact that the number 6 turns up—assuming that the dice are not loaded—simply cannot be the object of any causal account. It appears to be 'possible' in the sense of being consistent with our nomological knowledge. However, the conviction that the throw *must* 'necessarily' produce this result remains purely *a priori*" (RK, 126). The only scientific conclusion that we can draw about the result of any particular throw of the dice is that it "is consistent with our nomological knowledge." This is also the only scientific conclusion that we can draw about any individual particular event in the natural world. Weber asks us to consider the logical and methodological implications: "In the domain of the 'uninterpretable,' the *individual* event—the single throw of the dice, the splintering of the falling boulder—remains completely irrational in the following sense: we must be satisfied with the confirmation of nomological possibility—consistency with empirical generalizations" (RK, 126). If we desire to go beyond "confirmation of nomological possibility" as an account for an individual event "the most we can do is to describe a *plurality* of individual cases in terms of 'statements of relative frequency'" and, even that, only "given certain presuppositions" (RK, 126f.).

Weber asks us to contrast the single throw of the dice or the falling of the boulder with "the conduct of Friedrich II during the year 1756, that is to say, in a single, quite concrete situation" (RK, 127). When we combine the knowledge of Friedrich's behavior on the date and situation in question that is available to us from the historical sources, with "our own imagination, schooled in the experience of the everyday world, a genuinely positive *causal* interpretation in terms of 'motives' is produced" (RK, 126). This interpretation shows that the conduct of Friedrich:

> Is not only nomologically "possible," like the splintering of the boulder. It is also "teleologically" *rational*. Not in the sense that we can establish, as a result of the ascription of causes, a statement of *necessity*. But rather in the sense that his conduct has an "adequate cause." I.e., given certain intentions and (true or false) beliefs of the monarch and given also a rational action determined thereby, a "sufficient" motivation can be identified. (RK, 127)

In contrast to the throw of the dice and the falling boulder, the causal account of Friedrich's conduct can legitimately address the question: "Why did he act in this particular way on this particular occasion when he could have done differently?"While such a question is patently absurd with respect to the fall of a boulder or the roll of a dice, it becomes eminently rational with respect to human actions.

Weber's contrast of the particularities of a natural event and the behavior of an individual human being in particular circumstances establishes the fact that in the case of the behavior of the individual, "the 'possibility of interpretation' implies that 'interpretable' processes are 'more calculable' than 'non-interpretable' natural processes" (RK, 127). When the issue is looked at from this angle, it is the domain of socio-cultural events that becomes eminently "rational" and "calculable" and the domain of natural-physical events that becomes the eminently "incalculable" and "irrational." Weber notes:

> Because of its susceptibility to a meaningful *interpretation*—and to the degree that it is susceptible to this sort of interpretation—individual human conduct is in principle intrinsically less "irrational" than the individual natural event. Insofar as it is susceptible to a meaningful interpretation: if human conduct cannot be interpreted in this way, it is no different from the fall of the boulder from the cliff. (RK, 125)

We can say that whereas the causal interpretation of a natural event can go no further than providing a "nomological explanation" of the natural phenomenon, the causal interpretation of human action can provide an "interpretive understanding" of the observed phenomenon. Weber rejects the claim that human action is "incalculable" in the sense of not being amenable to interpretive understanding as being "the principle of the 'madman'" (RK, 125). The key point in Weber's critique and analysis of Knies's position is that human action is not "more 'irrational,' in the sense of 'incalculable' or inaccessible to casual explanation, than *any concrete* event as such" (RK, 192). In a very particular sense "whenever rational 'interpretation' is possible, action is withdrawn from the domain of the irrationality of the purely 'natural'" (RK, 192). In such a case not only is the action "withdrawn from the domain of the irrationality of the purely 'natural,'" it becomes eminently more rational. If rationality entails more than just making abstract generalizations about a general class of observed phenomena and also includes offering specific reasons for a specific event being as it is when it could have been otherwise, then it is clear that the soft sciences are more directly and consciously concerned with rationality than the hard sciences. Weber notes that this requires an inversion of Knies's "human freedom" thesis that posited a direct relationship between freedom and incalculability:

> Suppose that the extent to which an action is interpretable decreases. Put another way, suppose that the degree of its "incalculability" increases. To just that extent, we are inclined to *deny* that the actor has "freedom of will" (in the sense of "freedom of action"). In other words, these remarks show that even if there is some general relation between *"freedom" of action* (regardless of how

> this concept is understood) and the irrationality of the historical *event*, that relation is not the following: the conditions for their existence are equivalent, the existence or increase of one implies an increase in the other. [As the foregoing discussion has shown] clearly, just the contrary is the case. (RK, 128)

Weber's analysis of Knies's unpredictability thesis shows that human freedom makes human action more predictable than any natural event. Taken together with his analysis of Wundt's "creativity" thesis, Weber's critique of the intuitionist position shows human action to be more rational and predictable than events in the physical-natural domain. At the same time, Weber demonstrates that there is no scientific or logical warrant for investing causal agency solely in something that is unique to "human nature." He describes the common intellectual genealogy of Wundt's creativity thesis and Knies's freedom thesis: "Kant's conception of 'causality through freedom,' together with the various ramifications which grew out of this concept in the subsequent development of philosophical thought, is the philosophical archetype of all metaphysical 'culture' and 'personality' theories of this type" (RK, 118). Kant's formulation is nothing short of "grandiose and, in its logic, ruthlessly perspicuous" (RK, 119). But its grandiosity and perspicuousness notwithstanding, it is marked by "contradictions" and more importantly Kant's philosophy and the lesser theories that rose from it (including those of Wundt and Knies) are based on "a metaphysical belief" (RK, 118) that is completely divorced from empirical reality and thereby not amenable to scientific scrutiny. Hence, the intuitionist conception of causality reaches the same destination as the positivist conception through a different route.

NOTES

1. The term "iron cage" is the translation of *stahlrates Gehause* by Talcott Parsons; Kalberg has translated the phrase as "a steel-hard casing." I have inserted Parson's translation into text because of its widespread usage.

2. Here Weber is offering a semiotic account of the world of nature and suggesting that the process of semiosis is at work even in the animal kingdom. We will look at Peirce's semiotics in detail in chapters 3 and 4. For the time being it is worth noting that the biologist Jakob von Uexküll gave an account of "theoretical biology" that saw the process of semiosis taking place in the animal kingdom. For an account of this theory, see von Uexküll (2010). In hindsight, von Uexküll is considered a leading pioneer in the field of biosemiotics. Among others, Thomas Sebeok built on von Uexküll's insights and coined the term "zoosemiotics" to describe the process of semiosis taking place in the animal kingdom. For a readable account, see Sebeok (1994). Martin Krampen (1981) coined the term "phytosemiotics" and argued that the process of semiosis is at work in the plant kingdom. Kalevi Kull (2000) builds on Krampen's pioneering insights and offers additional evidence to support the claim that semiotics provides the best conceptual tools for studying the plant kingdom. Witzany (2005) goes even further and uses Peirce-like thinking to argue that biology should be conceived of as "an understanding social science." In another paper (2008) he uses semiotics to show that "the development and growth of plants depends on the success of complex communication processes. These communication processes are primarily sign-mediated interactions and not simply a mechanical exchange of

'information,' as that term has come to be understood (or misunderstood) in science" (Witzany, 2008, 39). Given his observation that since chemical molecules "may also function as signs" (Witzany, 2008, 60), a semiotics approach in chemistry is also hinted at.

Chapter Two

Weber's Conception of Causality

A Reconstruction

As the previous chapter illustrates, Weber's reflections on causality are dispersed in a number of different writings. The goal of the present chapter is to pull these reflections into one place from the perspective of the key issue that is of interest in this inquiry—the integration of the hard and soft sciences. The chapter will touch on three different issues in this regard. The first part will detail Weber's position that the scientific conception of causality is no different from the everyday commonsense conception. Weber uses the example of the causal account given by a mother spanking her child and a causal account of the historical significance of the Battle of Marathon to detail this point. The second part of the chapter will describe the philosophical resources that Weber used to construct his account of causality. In very explicit terms Weber rejects John Stuart Mill's conception of causality and builds his own conception in line with the insights of the legal theorist Johannes von Kries. For Weber the process used by jurists to determine the cause of a particular event is precisely the process by which the socio-cultural scientist provides a causal account of events in empirical reality. The third part of the chapter will detail Weber's observation that the soft sciences (in contrast to the hard sciences) "employ the category of causality in its full meaning" (RK, 196). This examination will show that the hard sciences can employ the category of causality only in a truncated and degenerate form. A scientific conception of "causality" is not theoretically, logically, or empirically possible unless the unique contribution of the soft science is incorporated into the conception—in exactly the same sense that a scientific understanding of "matter" is not theoretically, logically, or empirically possible without taking into account the unique contributions of quantum and relativity physics. Not taking the contributions of relativity and quantum physics into consideration and describing "matter" from a strictly Newtonian perspective will obviously produce a truncated and degenerate conception of "matter."

THE COMMON SENSE OF CAUSALITY

Weber notes that "the causal analysis of personal actions proceeds logically in exactly the same way as the causal analysis" of historically and culturally significant events (LCS, 177). This is the case whether we are analyzing our own action in first person terms or that of someone else in second or third person terms. This appears to be a remarkable claim on Weber's part—the logical structure of a causal account of an ordinary everyday event (for example, a mother spanking her child) is no different from a causal account offered by a historian for an event of world-historical significance (for example, the significance of the Battle of Marathon).Weber offers a detailed logical analysis of the two cases to support his claim. He begins with the case of a mother who has given "a solid cuff" (LCS, 177) to her child for misbehaving. After the event has occurred she might be called upon to explain the "cause" of her actions for a number of different reasons. The reasons could be internal (i.e., remorse on her part for her actions) or external (i.e., questions from the father). Reflecting on her actions she offers the following explanation to her husband: He knows (based on his familiarity with her typical disposition and behavior) that under ordinary circumstances she would not have spanked the child for misbehaving. Weber notes that here she is referring "to his 'empirical knowledge' of her 'usual motives'" (LCS, 178). But on this particular occasion she had had an argument with the cook earlier in the day, because of which she was unusually agitated. Had it not been for the argument with the cook, the "disciplinary procedure *would* not have been used at all or would not have been applied 'in that way'" (LCS, 177f.). In simple terms she is telling her husband that the misdeeds of the child were not the only factor that caused the spanking—the argument with the cook was an important factor. At the same time, the argument with the cook is also not the only causal factor for her action—if the child had been sleeping in bed and not misbehaving she would not have deliberately sought him out to spank him to vent her frustration with the cook. The cause of her action was the result of the agitation of the argument with the cook combined with the misbehavior of the child. Looking at the causal account of an everyday ordinary event provided by an ordinary actor, Weber makes the following observation: "The young woman would certainly be astounded if a logician showed her that she had made a causal 'imputation' just like an historian, that, to this end, she had made 'judgments of objective possibility' and had 'operated' with the category of 'adequate causation'. . . .Yet such is precisely and inevitably the case from the point of view of logic" (LCS, 178).

When the causal account of an ordinary event is analyzed using the tools of logic, we find that the causal account is composed of four different elements: (a) "empirical knowledge," (b) "adequate causation," (c) "judgments of objective possibility," and (d) "imputation." The terminology that is used is logical, technical, and precise—the reality to which this terminology refers is mundane, commonsense, and general. It is only by taking the time to carefully understand Weber's use of these four terms that we can understand what Weber means when he says that the there is no difference between the causal account of historically significant events and everyday (mundane, insignificant) events. Taken one by one we can offer the following definitions of these terms:

> *Empirical knowledge*: "Knowledge of certain known empirical rules, particularly those relating to the ways in which human beings are prone to react under given situations ('nomological knowledge')" (LCS, 174).

This empirical knowledge is a type of "nomological knowledge" in the sense that it is a generalization. But it is radically different from the way that "nomological knowledge" is understood and employed in the hard sciences because the generalization in the soft sciences is considered neither to be deterministic nor universal. It is the knowledge of how an individual, a group of individuals, or even the "average" individual will typically act under particular circumstances as opposed to any knowledge of "laws" that (allegedly) govern human action. While statistical regularities as understood in probability theory come closest to approximating this empirical knowledge, even they fall short of encompassing all the information contained in empirical reality about causal relationships. Empirical knowledge in the soft sciences is closely linked with the knowledge of motives. Weber defines motives as "a complex of subjective meaning which seem to the actor himself or to the observer an adequate ground for the conduct in question" (ES, 11). In the present example, empirical knowledge (the husband's knowledge of the mother's behavior under typical circumstances, or her "usual motives") leads to the conclusion that the "usual motives" are not an "adequate ground" (or "adequate cause") for the spanking that she gave—therefore an adequate cause has to be found.

> *Adequate causation*: Cases of adequate causation refer to those cases in which "the relationship of certain complexes of 'conditions' synthesized into a unity by . . . reflection and conceived as isolated" are causally linked "to an 'effect' that occurred" (LCS, 184).

Every effect is surrounded by an almost infinite number of conditions—in the present case a very partial list of these condition would be: the time the mother woke up in the morning, what she had for breakfast, the clothes she was wearing, the conversation she had with her husband the previous night, the events reported by the newspaper that day, the temperature outside, the season of the year, the agenda to be taken up by the German parliament that day, the price of rice in China during the day, the meeting between two tribal elders in Timbuktu, etc. Out of all of these conditions that preceded the event in question, the child's misbehavior and the argument with the cook are "synthesized into a unity" and "conceived as [the] isolated" (adequate) cause of the "effect" in question—to the exclusion of all other conditions that were also present. The two factors are labeled the adequate cause because these two factors are the ones "whose realization generally and significantly *favors* the outcome, that is, *increases the chances* of the outcome when compared to the other [co-existing antecedent] conditions" (Heidelberger, 2010, 250).

> *Judgments of (objective) possibility*: "the proposition regarding what would happen in the event of the exclusion or modification of certain conditions" (LCS, 173).

In light of empirical knowledge, one can offer a hypothesis of what might or might not have happened if certain parts of the event had been different. The mother makes the claim that based on his knowledge of her typical behavior, the husband knows that she would have acted differently if she had not been agitated by the cook. This proposition is a "judgment of objective possibility." These types of judgments allow us to engage in counterfactual reasoning. When we are able to conceptually manipulate the event in question we can better understand how the absence or presence of certain factors increased or decreased the probability of the event occurring. Counterfactual reasoning expands our intellectual capacities and helps us to gain keener insight into the factors that are at work. Weber notes: "In order to penetrate to the real causal interrelationships, *we construct unreal ones*" (LCS, 185f.)

> *Imputation*: "Where the *individuality* of a phenomenon is concerned, the question of causality is not a question of *laws* but of concrete causal *relationships*: it is not a question of the subsumption of the event under some general rubric as a representative case but of its imputation as a consequence of some constellation. It is in brief a *question of imputation*" (OSS, 78f.).

The cause of this mother spanking this child, on this day, at this place, is not and can never be "subsumed," "covered," or "governed" by any law that has been yet discovered or will ever be discovered. (And we may add that the cause of the spanking cannot be attributed to the "essence" or "true nature" of motherhood, even of those mothers who have been recently agitated for some reason.) The cause of any particular phenomenon can only be explained in terms of "concrete causal relationships" and these relationships can only be described in terms of "some constellation" of conditions in which the motives and intentions of a conscious individual are the causal agent(s)—and the description comes in the form of an "imputation." Weber's theory of causality can be called a "theory of causal imputation" and this is "a theory of attributing causality to agents" (Heidelberger, 2010, 247).

This example is an "explanation" that one person gives to another (and more importantly and interestingly to herself) for acting in a particular way on a particular occasion when one could have acted otherwise. It is clear that the account is full of logical elements and procedures. Weber notes that only "untrained sentiment" leads us to think that causal interpretation which seeks to understand "one's *own* action" (and we may add one's own feelings and thoughts) is free of logical elements and processes (LCS, 177). First person reflections can come in the form of the following questions: "What triggered those emotions in me at that time?" or "Why did I have those thoughts on that occasion?" or "Why did I do that in that situation?"The way we answer these questions determines how we attain first person understanding of our own thoughts, actions, and feelings. Weber notes we gain first person understanding by using the exact same logical elements and processes that are employed to give a second (or third) person explanation of our thoughts, actions and feelings to others—that is, empirical rules, judgments of possibility, adequate causation. These logical elements are pulled together in the form of an "imputation" that serves as an "explanation" (to ourself or to others). An "imputation" is a hypothesis that aspires to offer a "correct" causal account for an observed effect. Weber describes the contents of a "correct" causal interpretation in these terms:

> A correct causal interpretation of a concrete course of action is arrived at when the overt action and the motives have been correctly apprehended and at the same time their relation has become meaningfully comprehensible. A correct causal interpretation of typical action means that the process which is claimed to be typical is shown to be both adequately grasped on the level of meaning and at the same time the interpretation is to some degree causally adequate. (RK, 68)

According to Huff, there is remarkable similarity between Weber's conception of imputation and Peirce's notion of "abductive inference" (Huff, 1984, 68). Peirce's notion of abduction (as with Weber's notion of scientific inquiry) begins by observing an effect in empirical reality—more often than not the effect that draws our attention is a surprising, unexpected anomaly. After observing the effect, the attempt to understand the anomaly begins with trying to place it within the context of our knowledge of "empirical rules." On the basis of these empirical rules we construct "a *hypothetical* constellation of processes characterized by properties *a, b, c*" (Huff, 1984, 68). The assumption is that if this hypothetical actually obtained in reality then the surprising or anomalous characteristic of the phenomena would follow as a "matter of course" (Huff, 1984, 68). Another term Peirce uses for abduction is "retroduction"—the latter term captures the sense of moving from an observed effect via theoretical arrangement and rearrangement of known facts to a hypothesis about the cause. For Weber all understanding and explanation of ordinary human action is arrived at via retroduction/abduction. (This point will be discussed in detail in chapter 5.)

Weber repeatedly draws to our attention the fact that the very same logical elements and processes that are involved in providing a causal account for, and understanding the meaning of, mundane actions of ordinary individuals are at work when we attempt to understand the cause and meaning of socio-cultural phenomena in general. Commenting on Weber's position on this issue, Hekman notes:

> The form of causal analysis in everyday life and the sociocultural sciences is "of the same logical structure." This is the case because, in a broad sense, the social scientist and the everyday social actor employ causal analysis for the same purpose: to explain an action as meaningful. . . . The social scientist, unlike the social actor, can move beyond the actor's understanding and offer an explanation at a higher level of abstraction. But the use of causal analysis in everyday life and in the sociocultural sciences is similar because the subject matter in both cases is meaningful human action. (Hekman, 1979, 73)

Weber uses the example of the Battle of Marathon to illustrate this point. The battle took place in 490 BCE, pitting the Persians against the Greeks. Even though this battle was a relatively small affair in comparison with other battles between these parties, one of Weber's contemporaries (Eduard Meyer) imputes "world historical 'significance'" to this battle because it is of causal significance in "the development of western culture" (LCS, 171). Weber analyzes the logical process that leads up to this claim. Empirical knowledge shows that there were two cultural trends in Greek society in the early fifth century BCE. The first was "theocratic-religious culture, the beginnings of which lay in the mysteries and oracles" and the second one was what eventually came to be known as "the free Hellenic circle of ideas, oriented towards this world" (LCS, 171). Of the two trends, the theocratic-religious trend was probably stronger than the Hellenic circle trend at this point in Greek history. Empirical knowledge about the Persian Empire shows that when it conquered other nations it used the national religion of the conquered nation "as an instrument of domination" (LCS, 171)—by co-opting the religious symbols, traditions, and authorities of the conquered culture to legitimate Persian political authority. Empirical knowledge also shows that the Greeks won the battle. Based on this empirical knowledge, Meyer constructs a "judgment of possibility." He does this by conceiving "a few of the actual components" of empirical reality "as modified in a certain direction and then [asks] whether under the conditions which have been thus changed, the same effect . . . or some other effect 'would be expected'" (LCS, 171).

The most obvious part of the given scenario that can be changed is the fact that the Greeks won the battle. If the Persians had won they would have patronized the "theocratic-religious" trend and this would have probably led to the decisive defeat of the "Hellenic circle of ideas" trend. The battle was relatively minor in terms of settling matters between the Greeks and Persians (they would go on to fight many more and much bigger battles in the decades and centuries to come). But the Battle of Marathon was absolutely decisive in determining which of the two cultural trends within Greek society prevailed. Weber notes that in hindsight we see that the victory at Marathon proved to be an "indispensible 'precondition' of the development of the Attic fleet and thus the further development of the war of liberation, the salvation of independence of Hellenic culture, the positive stimulus of the beginnings of specifically western historiography, the full development of the drama and all that unique life of the mind which took place in this—by purely quantitative standards—miniature theater of world history" (LCS, 172). The combination of empirical knowledge, judgments of possibility, and adequate causation generates an imputation that the Battle of Marathon was historically significant—and of world historical significance. As noted above, the imputation is a hypothesis that has to be tested. The question naturally arises: How does one test such a hypothesis?

In the soft sciences it is not possible to design a laboratory experiment to test the validity of a hypothesis. One of the ways to test the hypothesis is by constructing hypothetical scenarios in which the original conditions of the event are modified and then comparing the outcome of the hypothetical scenario with the outcome of similar events in the past. In comparing the hypothetical constellation with the outcome of similar events in the past we can see which part of the event was "causally significant" and which "causally insignificant." In socio-cultural phenomenon it is often the case that the part of the event that is most "obviously" causally significant is not really that significant at all. For example, an uprising occurred in Berlin in 1848 that came to be called the "March Revolution." The events were triggered by two gunshots during the night, and the next day revolutionary fever gripped the entire city. Were these two shots as significant in "causing" the March Revolution as the victory at Marathon was in "causing" the emergence of Hellenic culture? We offer a judgment of possibility by constructing a hypothetical constellation in which the shots were not fired. Our empirical knowledge of the social, economic, and political conditions that have accompanied similar events in the past, combined with our knowledge of the circumstances surrounding this particular event, strongly suggests that the revolution would have happened even if the two shots had not been fired. The revolution may have started at a later date because of which the course that it took would have been different, but the shots themselves were "causally insignificant." The conditions surrounding the event were such that no particular happening that apparently "caused" the event—that is, two shots being fired, a street beggar being run over by the carriage of a government official, an argument triggered by the high-handedness of a police man—could be labeled "causally significant."

It is clear that causal interpretation in the soft sciences cannot be tested under controlled laboratory conditions. But this does not mean that there are no means of inductively testing them. Weber notes that causal interpretation in the soft sciences is tested against increased empirical knowledge of the event in question, of similar events in the past, or a similar new event happening in the future. In other words causal interpretation in the soft sciences is not tested under artificially created laboratory conditions that do not exist anywhere in the universe but against conditions in the actually lived and experienced universe. And the most important element in the actually lived and experienced universe against which the validity of the causal interpretation is measured is the investigative interests of the inquirer. To the degree that the causal interpretation helps the investigator better understand, address, and (what is probably most important) redress his investigative interests and concerns, his causal interpretation is valid. In putting forth his account of the different elements that go into a causal account of events and the way that the validity of this account is tested, Weber does not stand alone—he is relying on some reputable resources.[1]

JOHANNES VON KRIES AND THE LEGAL UNDERSTANDING OF CAUSALITY

Weber notes that legal philosophers have investigated the logical and methodological issues surrounding the concept of "causality" in greater depth and detail than any other science. This is due to the fact that "jurists must routinely deal with the problem of when someone can be said to 'cause' an event" (Hekman, 1979, 73). While there are some significant differences between jurisprudence and the soft sciences, in terms of the logical and methodological issues concerning the determination of the cause of a particular event there is no difference. When we look at the manner in which jurists pose the question of causality during criminal trials and the manner in which historians (and more generally socio-cultural scientists) pose the question of causality in socio-cultural phenomena, we find that the central question in both cases is exactly the same: "Under what circumstances can it be asserted that someone through his action has 'caused' a certain external effect, is purely a question of causation. And, indeed, this problem obviously has exactly the same logical structure as the problem of historical 'causality'" (LCS, 168). In addition to the central question being exactly the same, the empirical conditions in which the question has to be answered are also exactly the same: "We ask first, in common with juristic theory, how in general is the attribution of a concrete effect to an individual 'cause' possible and realizable in principle in view of the fact that in truth an *infinity* of causal factors have conditioned the occurrence of the individual 'event' and that indeed absolutely all of those individual factors were indispensible for the occurrence of the effect in its concrete form" (LCS, 169). The "cause" of an event would mean listing "an *infinity* of causal factors" if we were to offer an "objectively detached" description of the cause—such an account is neither logically possible nor scientifically meaningful. In the place of a logically absurd and scientifically meaningless conception of causality (and more generally of empirical reality) history and related socio-cultural disciplines are:

> Exclusively concerned with the causal explanation of those "elements" and "aspects" of the events in question which are of "general significance" and *hence* of historical [and socio-cultural] *interest* from general standpoints, exactly in the same way as the judge's deliberations take into account not the total individualized course of events of the case but rather those components of the events which are pertinent for subsumption under the legal norms. (LCS, 169f.)

At this point in the discussion Weber explains the point he is making with the aid of an example. Whether it is a judge trying to determine the cause of death in a murder case or an historian trying to determine the cause of death of Caesar, the investigator has to rule out innumerable factors that were part of the actual causal chain as being "causally insignificant" in the event and single out one factor (or a very small number of factors) as being "causally significant." In distinguishing the causally significant factors from the insignificant factors, the perspective of the investigator plays a decisive role. Weber notes that the "cause" could be very different if someone begins the investigation of the same event with a question that is different from the one asked by the judge or the historian. For the judge the decisive standpoint that determines which part of the causal chain is significant from his perspective is: "Whether the causal chain between the thrust and the death took place in such a form and the subjective attitude of the murderer and his relation to the deed was such that a certain norm of criminal law is applicable" (LCS, 170). In contrast to the judge, (assuming the murder took place in an airport) the security officer at the airport would be more interested in determining those causal factors in the event that made it possible for the murder to elude the security arrangements and smuggle the murder weapon into the place of the crime. The crime could have been (among other things) a crime of passion, an act of revenge, the result of temporary insanity, the unintended consequence of a botched robbery, etc. For the judge it is of critical importance to determine which of these subjective motives played a role in the murder. The security officer could not care less about the subjective motives of murderer that led to the crime. He is almost exclusively concerned with identifying the factors in the "lapse of security" that made it possible for the murder weapon to escape detection. The judge engages in the objectively observed fact of investigating an "object" in empirical reality (the murder) because of the "object's" relationship to the penal code that the judge has sworn an oath to uphold. For the security officer the same event becomes significant because it is a breach of security—irrespective of whether the breach violates any penal code. It is obvious that the character (or meaning) of the inquiry changes radically given the value-position of the investigator (valuing the penal code vs. valuing security measures).

The crucial role that values play in shaping the character and conclusions of an inquiry can be summarily illustrated by looking at the final judgments of the judge and the security officer. "Objectively" speaking the intent of the murderer is as much a "cause" of the murder as the lapse of security (along with innumerable other causes). But in this particular case we find both the judge and the security official identifying one factor (or a small number of factors) as "causally significant" to "the exclusion of an infinity of components of a real action as 'causally irrelevant'" (LCS, 171). At this point Weber brings us face to face with the crux of the logical and methodological issues surrounding causal imputation. He notes: "Our real problem is . . . by which logical operations do we acquire the insight and how can we demonstratively establish *that* such a causal relation exists between those 'essential' components of the effects and certain components among the infinity of determining [causal] factors" (LCS, 171). For Weber the only way to separate the "causally significant" factors from the "causally insignificant" factors is with reference to the investigative concerns (or "subjective" values) of the investigator.

From the discussion in the previous chapter it is obvious that neither the positivist conception of the "laws of nature" nor the intuitionist conception of "human nature" can help us to distinguish the "causally significant" from the "causally insignificant" factors in an event. On the surface John Stuart Mill's conception of causality appears to be promising. In contrast to the completely naïve claim of the positivists that every individual event in empirical reality is the effect of a universal, abstract law, Mill noted that our everyday experience and observations show that it is rarely (if ever) the case that a particular event is preceded by a single causal condition. Careful consideration shows that every particular event is the outcome of numerous antecedent conditions. Mill argues that some of these antecedent conditions work in conjunction to favor the occurrence of the event, and some of the antecedent conditions work in conjunction to obstruct its occurrence. An event occurs (or an effect appears in reality) at that particular point in time when the concurrence of the antecedent conditions favoring the event is greater than the concurrence of the antecedent conditions obstructing it. This conception of causality attracted the attention of many legal theorists because "its empirical nature and high flexibility made it easily applicable to cases in everyday life" (Heidelberger, 2010, 246). But Mill's notion of causality is of very limited (if any) use in legal theory. From the perspective of legal theory, the most important point that is left unaddressed by Mill is the following: "Which of the conditions is to be regarded as decisive for attributing responsibility of an event to someone? If the cause of an event is a whole set of conditions including not only a particular human act but also a whole lot of other circumstances, how can an agent be held responsible for his or her deed?" (Heidelberger, 2010, 247). With respect to the antecedent conditions that favored the occurrence of the event, Mill's theory contains no objective criteria that could be used to distinguish causally significant antecedent conditions from causally insignificant antecedent conditions. Following Weber's example, let us consider an event from the perspective of Mill's logic. An individual has stabbed someone to death. From the perspective of Mill's "antecedent conditions," the murderer plunging a knife into the victim is as much a cause for the victim's death as the mother who gave birth to the murderer (or the mother who gave birth to the victim for that matter). From this perspective, the birth of the victim and perpetrator are as much antecedent conditions as the intention of the murderer to kill the victim. In Mill's logic labeling the intention of murderer as "causally significant" is a purely arbitrary judgment.

In addition to this practical shortcoming, Weber critiques Mill's conception on logical grounds. Whatever the value of Mill's conception of favoring and obstructing antecedent conditions in studying purely mechanical phenomena, "it is . . . certain no rigorous causal analysis, even in history, can accept this anthropomorphism" (LCS, 186). In logical terms, we cannot categorize the antecedent conditions of an effect either favoring or opposing the effect:

> Rather it is to be emphasized once and for all that a concrete result cannot be viewed as the product of a struggle of certain causes favoring it and other causes opposing it. The situation must, instead, be seen as follows: the totality of *all* the conditions back to which the causal chain from the "effect" leads had to "act jointly" in a certain way and in no other for the concrete effect to be realized. In other words, the appearance of the result is, for every causally working empirical science, determined not just from a certain moment but "from eternity." (LCS, 187)

From the point of view of logic, every single particular event in empirical reality is the outcome of a causal chain that extends across all known space and all the way back into past eternity.

In his critique of Mill and in the articulation of his alternative, Weber relies heavily on the work of Johannes von Kries. Weber notes that von Kries has not only "shown the contrast between his theory and John Stuart Mill's . . . in a way which is entirely convincing to me" but also that von Kries has "convincingly criticized" Mill's conception of causality (LCS, 186). Toward the end of the essay, when he begins to summarize the salient features of his own conception of causality, Weber inserts the following footnote: "I scarcely [need to] mention the extent to which here again, as in so much of the preceding argument, I am 'plundering' von Kries' ideas" (LCS, 186). For Weber, von Kries is the one thinker whose conception of causality has most satisfactorily solved the logical problem that is common to both legal theory and socio-cultural inquiry—given the fact that every effect is the result of innumerable causal factors, how do we distinguish the causally significant factor(s) from the causally insignificant factors?

At this point it is worth recalling that in the first section of the present chapter Weber argued that the logical process by which a socio-cultural investigator determines the cause of a particular event is exactly the same as the logical process by which ordinary individuals attain self-understanding of their own actions and explain their actions to others. Weber is not the only one who recognized that the commonsense understanding of causality makes an important contribution to the scientific conception of causality. The initial insight that led von Kries to formulate his conception of causality was based on deriving the legal understanding of causality from the commonsense understanding: "Johannes von Kries' original intuition . . . was that there *must* be an *objective* reason for singling out in 'common parlance' just one factor from the set of antecedent conditions as *the* cause" (Heidelberger, 2010, 248). In observing human action it is obvious that human beings routinely single out one factor (or a very small number of factors) as *the* cause for particular concrete events—even though logically speaking the causal factors in any particular event are innumerable. In most cases, commonsense causal analysis proves to be correct in light of future experience. Von Kries's original insight was that a logic is at work in the commonsense conception of causality and this conception could be of use to legal theorists. Without introducing technical language "that would be available only to legal experts" von Kries "claims to have shown that (and how) our everyday causal and legal intuitions in attributive contexts are guided by an objective criterion that can be made explicit" (Heidelberger, 2010, 250).

Von Kries's starting point in making the logic that underpins ordinary everyday intuitions explicit is his range theory of probability. Before turning his attention to the problematic situation in the area jurisprudence, von Kries had used his probability theory in a variety of other areas to resolve some of the difficult questions of the day. Fully aware of the fact that any given effect is the result of many different causal factors, von Kries begins with the idea that "there is usually only one condition that *generally* increases the objective *possibility* of an effect in a significant way, that is, that favors an effect in *general*" (Heidelberger, 2010, 250). The term he used to describe this condition is "adequate cause"—"a necessary antecedent condition of an event whose realization generally and significantly *favors* the outcome, that is, *increases the chances* of the outcome when compared to other conditions" (Heidelberger, 2010, 250). When compared with all the other causal factors that lead up to a particular event, the "adequate cause" is the one factor that most significantly enhances the tendency or "objective possibility" of the event occurring. Von Kries uses the example of the roll of dice to illustrate his point. Our ordinary experience shows that when a dice is rolled the chances are 1/6 the number 6 will turn up. But if after repeated throws we find that a 6 is turning up on every third throw, we will look for a cause (or the favoring condition) that would explain the observed effect (which is a departure from the expected results). Heidelberger summarizes the relationship between the dice example and determining adequate cause in sociocultural phenomena in these words: "Since the realm of social phenomena is also characterized by mass repetitions of similar cases (like the game of dice), we are justified to use the same terminology for the social sciences" (Heidelberger, 2010, 250). After observing "mass repetition of similar cases," we develop a conception of "empirical rules" or "causal generalizations" that describe the general pattern of conduct, events, and expectations. When there are departures and deviations from these expectations, we determine the cause of the departure by identifying the factor that (in light of and in line with our experience) is most likely to have caused the deviation and call it the "cause"—in logical terms it should be called the "adequate cause." Here the causal account does not depend so much on knowledge of the "universal laws of nature" but on knowledge of typical motives, trends, and conduct of typical human begins under typical conditions.

THE FULL MEANING OF CAUSALITY

We have already noted that since the causal account offered by the soft sciences (i.e., the "imputation") cannot be tested in the controlled conditions of a laboratory, it appears that the understanding of causality as employed by these sciences is deficient when compared to the hard sciences. Weber argues that the exact opposite is the case because the soft sciences "employ the category of causality in its full meaning" (RK, 196). He details his position in these words:

> Those empirical disciplines which employ the category of causality and investigate the *qualities* of reality—history and every "science of culture" of any sort belong to this group of disciplines—invariably employ the category of causality in its full meaning. They conceive the circumstances and changes within concrete reality as "effected" and as "effective." In part, they attempt to establish "causal generalizations" by abstracting from the concrete properties of a complex. In part, they attempt to "explain" concrete causal complexes on the basis of these generalizations. (RK, 196)

If we unpack the contents of this passage, we come up with three points of contrast between the soft and hard sciences with respect to the notion of causality:

Table 2.1.

Soft Sciences	Hard Sciences
Reflexive relationship between effect and cause	Linear cause-and-effect link
Causal variation	Causal equivalence
Empirical phenomena as ultimate reality	Heuristic abstracts as ultimate reality

When we look at each of these three distinguishing features more closely we see how the soft sciences employ the category of causality "in its full meaning." First and foremost the soft sciences "conceive the circumstances and changes within concrete reality as 'effects' and as 'effective'" (RK, 196). Simply put, the soft sciences posit a dynamic and reflexive relationship between effect and cause as opposed to the static and linear link assumed by the hard sciences. The difference between the two positions can be illustrated by comparing the Protestant ethic thesis as understood by Weber with the law of gravity as understood by a typical physicist. Weber's analysis shows the "'spirit' of capitalism" to be the effect of the "Protestant ethic"—at one point in history. In late sixteenth- and early seventeenth-century Western Europe, the "Protestant ethic" was the cause for the emergence of the "'spirit' of capitalism" and also the determining factor that shaped its character. Weber goes on to show that at a later point in history the "'spirit' of capitalism" becomes a cause for effecting fundamental changes in the "Protestant ethic." In broad terms, Weber's analysis shows that religious factors were "causally significant" for the emergence of fundamental and far-reaching changes in the economic sphere of culture. These changes led to the emergence of modern capitalism. Once modern capitalism came into existence, in its turn it became "causally significant" in causing far-reaching and fundamental changes in the religious sphere of culture. Combined with other developments in modern culture, the economic and technical conditions engendered by modern capitalism are such that "the imitation of the life of Buddha, Jesus, or Francis seem condemned to failure for purely external reasons"

(RRW, 357). Weber sees the effect-cause-effect dynamic permeating all socio-cultural phenomena. The relationship is so intimate and dynamic that it is impossible to reach an Archimedean point in the analysis and identify a cause of any event in empirical reality. For Weber the effect everywhere and always reacts back on the cause and becomes a cause of effecting changes in the cause.

Stated in these terms the contrast with the hard sciences is clear. For the typical physical-natural scientist, the cause of the apple falling from the tree is something called the "law of gravity."At no time and in no place in the known universe can the typical scientist imagine any "falling apple" (in any sense of the word) ever causing any change in the "law of gravity." For the typical scientist, the relationship between effect and cause is static and linear. Events in empirical reality have no effect, influence, or impact on the "laws of nature" that (allegedly) cause these events, in any conceivable sense of the word "effect." The hard sciences "conceive the circumstances and changes within concrete reality" as "effects" which have no capacity to be causes that engender any changes in their respective "causes." The first distinguishing characteristic of the category of causality in the soft sciences in contrast to the hard sciences is that for the soft sciences, events that are "effects" in empirical reality at one point in time can become "causes" at another point in time. This is so much the case that the original effect can become the cause of far-reaching changes in the character of the original cause.

The second distinguishing characteristic of causality as it is used in the soft sciences is closely related to the first—the hard sciences are primarily interested in causal equivalence while the soft sciences are interested in causal variation. Weber notes that when looking at socio-cultural phenomena the investigator must keep in mind that "from the point of our 'conception' of historical reality, the causal course of events is susceptible to intentional and extensional *variations in meaning*" (RK, 103). The example of "economic profit" illustrates this point. As we saw in the previous chapter different meaning has been conferred upon "economic profit" at different times in Western history. From the point of view of medieval Catholic piety and economic traditionalism, economic profit is a necessary evil. From the point of view of the Protestant ascetic, it is a sign of salvation. For the modern capitalist economic profit is an end in itself. For Weber variation in meaning is a necessary but insufficient condition for making a phenomenon worthy of scientific analysis. What interests him is "the causal course of events" responsible for "the intentional and extensional *variations in meaning*" of economic profit that has been observed in empirical reality. Weber notes: "It is precisely [the] cases of historical *variations in meaning* that arouse our historical interest to the greatest extent. It follows that the specific historical task of the [socio-]cultural sciences is profoundly antithetical to the aims of all disciplines which attempt to reduce phenomena to causal equivalences"(RK, 104).

In contrast to the focus on causal variation in the soft sciences, the hard sciences are exclusively concerned with expressing their conclusions in the form of causal equivalences—that is, mathematical equations such as F=Ma, $E=MC^2$, PV=nRT. If the soft sciences were to mimic this approach they would end up with something like this:

- economic profit=necessary evil (the medieval meaning)
- economic profit=sign of salvation (the early modern meaning)
- economic profit=end in itself (the modern meaning)

For Weber such statements are crucial and indispensible means for moving toward the type of knowledge that the soft sciences aspire to attain. But they are utter nonsense if they are treated as conclusions of socio-cultural inquiry. Weber's conception of the soft sciences aspires to identify those factors that caused the *Sinn* (meaning) of "economic profit" to change from necessary evil to sign of salvation to end in itself during the course of Western history. Consequently, the task of soft sciences is "profoundly antithetical to the aims" of those sciences that are primarily interested in causal equivalences. The second distinguishing feature of causality in the soft sciences is that causal variation is of primary interest not causal equivalence.

The third distinguishing feature is closely related to the second. Weber notes that when we look at empirical reality from the perspective of the soft sciences we find that nowhere is any rule the cause of social action. In other words, nowhere is human action exclusively determined by some "law of nature" or "something inherent in human nature." Action is everywhere and always caused by the combination of a rule and a subjective intention to abide by the rule on the part of an actor. To illustrate this point Weber uses an example:

> For example, it is obviously not the conventional rule of greeting that tips my hat when I meet an acquaintance. On the contrary, my hand does it. But what is causally responsible for this? I may merely be in the "habit" of following such a "rule." In addition, I may know from experience that my acquaintance would regard my failure to greet him as a lapse of propriety. The result would be unpleasant. In this case, therefore, the action is a consequence of a "utilitarian" calculation. Or, finally, I may act on the belief that it is "not proper" to me to disregard a harmless "conventional rule" that is universally observed unless there is some compelling reason to do so. Therefore I act on the basis of my "idea of a norm." (CS, 108)

There can be a variety of reasons why an individual tips his hat upon seeing an acquaintance—out of habituation, due to utilitarian concerns, or conscious adherence to social norms. While the reasons are varied (and some reasons may even conflict with others), at no point and under no circumstances can the "rule" to tip one's hat be labeled as the "cause" of the action. A causal account of this particular event will obviously take advantage of the empirical rule that in certain European cultures people tip their hat as a show of respect. But in any particular case this rule cannot be called the "cause" when an individual is actually observed tipping his hat. Nor can it be labeled as the "cause" of the general class of events that come under the heading "tipping hats."

This example allows us to better understand what Weber means when he says that the soft sciences "employ the category of causality in its full meaning." Upon observing an event in empirical reality (let's say the tipping of a hat) we can make the following inferences. Firstly, at one point in time the event was possible (or a potentiality) that became actual at a later point in time. Secondly, the fact that we have observed similar events in the past suggests that there is some general rule (our "nomological" knowledge) that applies to the observed event. Thirdly, given the fact that only some people tip their hats under certain conditions, we know that a subjective motive plays a role in determining if the hat is tipped or not. When an event is actually observed in empirical reality, the hard sciences can only say that the occurrence of the event is "not in contradiction to our nomological knowledge." In the present case, the hard scientist cannot say anything more than "the tipping of hats when passing by an acquaintance does not contradict our knowledge of behavior patterns in Western culture." This is the same as the case when an apple falls from a tree. The falling of the apple is not in contradiction with the known laws of gravity or the "laws of nature." The laws of gravity, nature, cultural norms, and so forth are no more the cause of any observed event than the fraction 1/6 is the cause of a particular number turning up on a throw of the dice. The law of gravity describes the motion of the apple after it has begun to fall to the earth and 1/6 is a ratio that describes an expectant result—in neither case do we have the "cause" of the event. In the case of the falling apple, the logically correct statement would be: "The falling of the apple does not contradict our knowledge of the laws of gravity." When a person is observed tipping his hat, the causal account goes beyond the relatively uninteresting statement that the tipping of hats "does not contradict our knowledge of human behavior in Western culture." The soft sciences use nomological knowledge as a means toward a larger end—that is, construct an interpretive account that offers a testable hypothesis about the subjective motives of the actor that caused the objectively observed event.

Here the difference between the hard and soft sciences becomes stark. The soft sciences offer a type of causal account that the hard sciences cannot even theoretically imagine, let alone provide. The fact that the former cannot reach the third level of analysis, combined with the fact that most physical-natural scientists are very bad philosophers, leads to the oft-repeated and (totally fantastic) claim that the "laws of nature" are the "cause" of events in empirical reality (this is the mindless parroting of eighteenth-century positivist dogma). Under these conditions the term "reason" comes to be associated with the immutable, impersonal, and universal "laws of nature" that are allegedly the "cause" of not just particular events and processes in empirical reality but of empirical reality itself. In contrast, the soft sciences locate the "reason" of events in empirical reality in a motive or an intention on the part of a conscious and rational actor. For the socio-cultural scientists, the empirical rules, causal generalizations, nomological knowledge, etc. (i.e., the "laws of nature"), are neither the "cause" nor the "reason" for observed phenomena—they are indispensible heuristic tools needed to attain scientific knowledge of the "cause" and "reason" of observed phenomena. In the soft sciences the discovery of laws is not the end of scientific inquiry—it is a necessary means for attaining the end. The end is a causal interpretation of empirical phenomena. The limitations of "laws" in scientific inquiry are made obvious by looking at one of the soft sciences that is closest to the hard sciences—economics. Weber notes:

> Even in the case of all so-called "economic laws" without exception, we are concerned here not with "laws" in the narrower exact science sense, but with *adequate* casual relationships expressed in rules and with the category of "objective possibility." The establishment of such regularities is not the *end* but rather the *means* of knowledge. . . . In the cultural sciences, the knowledge of the universal or general is never valuable in itself. (OSS, 80)

The fact that the discovery of "laws" (and other generalizations) is merely an intermediary stage in scientific inquiry is made even more obvious when one looks at a science that is more solidly in the domain of soft sciences than economics—history:

> Where the *individuality* of a phenomenon is concerned, the question of causality is not a question of *laws* but of concrete causal *relationships*: it is not a question of the subsumption of the event under some general rubric as a representative case but of its imputation as a consequence of some configuration. It is in brief a *question of imputation.* Whenever the causal explanation of a "cultural phenomenon"—an "historical individual"—is under consideration, the knowledge of causal *laws* is not the *end* of the investigation but only a *means*. (OSS, 78f.)

To restate this point in the words that Weber uses, a causal account in the soft sciences is composed in part of causal generalizations or empirical rules abstracted "from the concrete properties of a complex" and in part of an explanation of the "concrete causal complexes on the basis of these generalizations" (RK, 196). In other words, the point at which the hard sciences end their inquiry (the discovery of generalizations, empirical rules, or "laws of nature") is but the intermediate stage of inquiry in the soft sciences.

This section began by citing a passage from Weber in which he notes that the soft sciences "employ the category of causality in its full meaning." An important theme in the subsequent discussion was the fact that the soft sciences are able to achieve this because they do not make the logical mistake of the positivists that conflates abstract, heuristic devices called the "laws of nature" with causal connections in empirical reality. At the same time Weber does not attribute causal agency to something that is unique to "human nature." Avoiding these mistakes allows Weber to see elements of conceptual and empirical reality that will forever elude the positivists and opens up the conclusions of his results to tests of logical and empirical validity that will forever elude the intuitionists. Weber's analysis of the manner in which causality is employed in the soft sciences teaches us that every causal account of empirical phenomena has to be triadic in character—whereas for the positivists it remains (at best) dyadic. The latter see empirical reality as a necessary effect of immutable, impersonal, universal laws of nature. Weber sees empirical reality as the objectively possible effect that results from an observable pattern of action (that can be expressed as an empirical rule) that is catalyzed by a conscious (but more often semiconscious) motive that seeks to attain particular ends. In Weber's triadic concept of causality, motives play the catalytic role of "causal agent."

NOTES

1. The categorical demand for "experimental proof" or "experimental evidence" to validate a hypothesis may have been justified during the relatively primitive and underdeveloped stage of the evolution of science. But the limitations of this demand should be clear to all practitioners of science by now in light of the work of Kurt Gödel. In 1931 he published two incompleteness theorems. Roger Penrose describes the first of these theorems as the "most important theorem in mathematical logic of all time" (Penrose, 1994, 14). Simply stated, this theorem shows that no system of mathematics can be (a) finite, (b) complete, and (c) consistent. The second incompleteness theorem showed (in simple terms) that given a particular mathematical system, there exist meaningful statements that can neither be proven nor disproven from within that particular system. Alfred Adler notes that Gödel's work

> proves that there exist meaningful mathematical statements that are neither provable nor disprovable, now or ever—neither provable nor disprovable, not simply because human thought or knowledge is insufficiently advanced but because the very nature of logic renders them incapable of resolution, no matter how long the human race survives or how wise it becomes. There is no way to escape this conundrum. It is not a question of sophistry of any kind. . . . The philosophical implications are devastating. (Adler, 1991, 439)

In light of Gödel's findings, the following (well-known) observation by Weber should not sound so outlandish: "An 'understanding' ('*Verständnis*') of human behavior achieved through interpretation contains in varying degrees, above all, a specific qualitative 'self-evidence' ('*Evidenz*')" (CIS, 151). Weber goes on to note that this does not exempt such statements from facing tests of validity; it is just that there are ways to test the validity of such statements other than laboratory experiments.

Chapter Three

The Significance of Concept Formation

The second point on which Weber sees no difference between the soft and hard sciences is "the significance of concept formation." Weber describes scientific knowledge as "the provision of concepts and judgments which are neither empirical reality nor reproductions of it but which facilitate its analytical ordering in a valid manner" (OSS, 111). For Weber the "concept" is embedded in the very definition of science. Since both the soft and hard sciences are branches of science, concept formation plays an equally important role in both branches. While causal interpretation is the end that scientific inquiry aims for, "concepts and judgments" that "facilitate [the] analytical ordering [of reality] in a valid manner" are the means used to attain the end. For Weber conceptual knowledge of empirical reality is a unique characteristic of scientific knowledge and one "will seek in vain for other truth to take the place of science" in this respect (OSS, 111). Combined with his observation that scientific inquiry is unique because its ultimate aim is to arrive at causal interpretation of events, we can summarize that scientific knowledge has unique characteristics when compared with aesthetic and ethical knowledge, (a) in terms of the end that it aims for (i.e., causal interpretation) and (b) in terms of the means used to attain that end (i.e., conceptual mastery of empirical reality). Having looked at the end that scientific inquiry aims for in some detail in the previous two chapters, this chapter and the next one will look at the means used to pursue the end.

For Weber, inquiry in the soft sciences is no less scientific than inquiry in the hard sciences. The distinction between the soft and hard sciences cannot be on the basis of the use (or nonuse) of concepts because, by Weber's definition, all scientific inquiry employs concepts. The difference comes in the type of concept that is employed. The name that Weber gives to the concepts used in socio-cultural inquiry is "ideal-type" and the name he gives to the type of concepts used in physical-natural inquiry is "generic concept" (OSS, 100). Leaving a detailed comparison of the two types of concept for later, the discussion will begin by offering examples of the types of objects over which conceptual mastery is made possible by the ideal-type. Once again we will turn to *The Protestant Ethic and the Spirit of Capitalism* for the examples—the "'spirit' of capitalism" and the "Protestant ethic" themselves being examples. Both the "'spirit' of capitalism" and the "Protestant ethic" are "unique individuals." The unique individual is characterized by certain qualities, sentiments, and value judgments. The ideal-type makes it possible to conceptually master the "subjectively" held aesthetic and ethical commitments of the unique individual based on the study of "objectively" observable patterns of actions (or habits). In addition to facilitating conceptual mastery over unique individuals, the ideal-type also brings to light the role of the investigator in the inquiry. The manner in which the investigator constructs the ideal-type (the objective part of his inquiry) is significantly determined by the research interests of the investigator (the subjective part of the inquiry). Consequently, in addition to facilitating conceptual mastery that makes it possible to capture unique individuals in empirical reality, the ideal-type also makes it possible to capture the objective and subjective sides of the inquiry—revealing that both the object that is being studied and the investigation that is being carried out have objective and subjective sides. Very much like Weber's conception of causality that establishes a dynamic and reflexive relationship between effect and cause, Weber's conception of the ideal-type establishes a reflexive relationship between subject and object. By making it possible to conceptualize the objective and subjective sides of the object that is being studied as well as the subjective and objective elements that go into the study itself, Weber lays the groundwork for showing that the soft sciences employ the category of concept "in its full meaning"—just as they employ the category of causality "in its full meaning."

After detailing Weber's description of the ideal-type, the discussion will summarize his critique of the positivists and intuitionists on the issue of concept formation. The positivist argued that the type of concepts used by the hard sciences had produced the most precise and objective scientific knowledge. In order for the soft sciences to attain the same level of objectivity, they would have to mimic the hard sciences in terms of the types of concepts used. The intuitionists argued that since the object studied by the soft sciences has a unique characteristic (i.e., it belongs to the domain of *Geist*), the means of accessing the object are also unique (i.e., intuition, empathy, or some type of "reliving" of the experience). For the intuitionists, socio-cultural inquiry could dispense with the use of concepts altogether. They went on to argue that the "objectivity" of socio-cultural inquiry is compromised if concepts are employed. Weber's analysis shows the logical and commonsensical flaws in both these positions.

THE "'SPIRIT' OF CAPITALISM" AND "PROTESTANT ETHIC" AS IDEAL-TYPES

The terms "spirit," "capitalism," "Protestant," and "ethic" are widely used in ordinary language and different meanings are attached to them. In his investigation, Weber has a very specific meaning (definition) in mind when he uses these terms—and this meaning is eventually presented in the form of an ideal-type. The ideal-type "'spirit' of capitalism" does not refer to a universal phenomenon, it refers to a particular, unique historical phenomenon that emerged at a particular time, in a particular place in history and then spread to other places at other times. The soft sciences are particularly interested in the study of unique individuals. The contrast with the hard sciences on this point can be illustrated by looking at the distinction between nomothetic and ideographic knowledge as described by Wilhelm Windelband. His insight allows us to see that "for the purpose of scientific analysis, any given lump of coal would be considered interchangeable with other individuals of its type, whereas Goethe would not be similarly interchangeable with other poets or historical figures of his age" (Ciaffa, 1998, 53). The hard scientist is interested only in those lumps of coal that conform to the general definition of "coal." The practitioner of the soft sciences is interested (generally speaking) in Goethe to the degree that Goethe is different from the typical "poet." In short, the hard sciences pursue "nomothetic" knowledge (knowledge of generals), while the soft sciences pursue "ideographic" knowledge (knowledge of individuals). The following passage explicitly identifies the "'spirit' of capitalism" as an "individual." Weber notes:

> The title of the study uses a concept that sounds rather intimidating: the "'spirit' of capitalism." What should be understood by it? . . . If one can discover at all an object for which the phrase *spirit of capitalism* is meaningful, then it can only be [an historical individual].[1] Such a singular entity is nothing more than a complex of relationships in historical reality. We join them together, from the vantage point of their *cultural significance*, into a conceptual unity. (PESC, 13)

From this passage it is clear that the "'spirit' of capitalism" is a unique historical individual. This is another way of repeating something that has been noted since the first chapter. Weber's investigation concerns the origins of a very particular, unique type of capitalism—not an inquiry into the birth of capitalism in a general sense.

Just as it is important to understand that Weber is referring to a unique historical individual when he uses the term the "'spirit' of capitalism," it is important to understand what Weber means by "individual." To begin with, an individual is "a complex of relationships in historical reality" (PESC, 13). This rules out an atomistic conception of the individual in which the "individual" refers to some undifferentiated, homogenous entity set apart from everything else by clear and distinct borders. The heterogeneous complex of relationships that collectively constitute the unique individual, for their part, are not a given in history, nature, or reality—they are joined together "into a conceptual unity" by an investigator who finds the unique individual to be culturally significant for some reason. The formulation of the conceptual unity is neither a purely subjective exercise on the part of the investigator nor is the investigator able to do this from some objectively detached perspective. The "conceptual unity" is constructed from the culturally significant perspective of an observer/investigator who finds the phenomenon worth studying for some reason.

This means that the vantage point of the investigator plays a critical role in the construction of the ideal-type. Weber explicitly acknowledges that the "'spirit' of capitalism" could be investigated from vantage points other than the one he has adopted, and in such a case the ideal-type would be different than the one that he uses: "It must be recognized that these vantage points (which are still to be discussed) by no means constitute the only ones possible in reference to which the historical case under consideration can be analyzed. Other vantage points would identify other features of our historical cases as 'essential,' as is true with every historical case" (PESC, 13). Consequently, "it follows unequivocally" that the "'spirit' of capitalism by no means necessarily can or must correspond to *that which we* will note as essential in our exegesis here" (PESC, 13). The fact that the "'spirit' of capitalism" (even though it is a unique individual) can be described in a variety of ways does not make any sense without first rejecting the notion that the "individual" or the "individual uniqueness" of a phenomenon is some homogeneous, undifferentiated entity (often called the "essence" or "true nature" of the phenomenon). For Weber, the individual is composed of an entire constellation of constituent components and the different components are brought to light (or can only be brought to light) by studying the unique individual from different perspectives.

Weber describes another set of the logical issues involved in formulating a clear concept of an individual—with specific reference to the "'spirit' of capitalism" as an individual:

> Such a historical concept, however, cannot be defined according to how it is "demarcated" vis-à-vis other concepts (*genus proximum, differentia specificia*). This holds if only because the concept denotes a phenomenon that is of

> qualitative importance as a consequence of its individual *uniqueness*. Moreover, this concept must be gradually *put together* from its single component parts, each of which is taken out of historical reality. (PESC, 13)

In this passage, Weber links "individual uniqueness" with "qualitative importance." In other words, the ideal-type will conceptualize certain qualities in the phenomenon that lend the phenomenon its uniqueness. This qualitative character of the phenomenon cannot be captured by any definition that categorizes the phenomenon as an instance of a general type—it can only be captured by "gradually put[ting] together" the "single component parts." The component parts are gathered from closely observing the unique phenomenon and discerning some regular patterns of action (or habits) that it displays. In short, the ideal-type aspires to capture the habitual regularity (in terms of attributes and actions) of the unique individual. Weber goes on to note that: "The final formation of the concept cannot appear at the beginning of the investigation; rather, it must stand at its *conclusion*" (PESC, 13). The bulk of Weber's inquiry is dedicated to describing the constellation of relationships that constitutes the unique individual called the "'spirit' of capitalism" and the unique individual called the "Protestant ethic"—and in the course of the discussion building a case to show that the latter is the "cause" of the former.

The question naturally emerges that if there is no concept of the "'spirit' of capitalism" at the beginning of the inquiry, how will the inquiry actually begin. There has to be some provisional understanding of the term "'spirit' of capitalism" at the beginning of the inquiry if the inquiry is to make any sense before it reaches its conclusion. The way Weber begins his inquiry illustrates that we can attain a partial understanding of the "'spirit' of capitalism" without having to resort to conceptual definitions—by way of an example from everyday ordinary life. Weber begins the investigation by drawing the reader's attention to a concrete phenomenon in empirical reality in which he sees the "'spirit' of capitalism" embodied. Looking at Benjamin Franklin's print shop in colonial America, Weber sees a peculiar mode of organization and attitude toward work. The way that the print shop was organized, and more importantly the ethos that determined the way that the print shop was run, serve as an example of the "'spirit' of capitalism." So instead of beginning with a conceptual definition of the "'spirit' of capitalism," Weber's investigation begins with observed effects in experienced reality (often occurring in the form of an observable pattern of action) serving as an example of a culturally significant phenomenon.

One of the "individual parts . . . from historical reality" that constitute the "'spirit' of capitalism" is its "cause." As noted in the previous chapter, Weber identifies the "Protestant ethic" as the "cause" of the "'spirit' of capitalism." Just as the effect is a unique, historical individual, the cause is also a unique, historical individual. As is the case with the "'spirit' of capitalism,"

Weber does not give a definition of the "Protestant ethic" at the beginning of the inquiry—he gives a number of examples. Just as Ben Franklin's print shop is an example of the "'spirit' of capitalism," Richard Baxter's (d. 1691) sermons are an example of the "Protestant ethic." There is no difference between the "'spirit' of capitalism" and the "Protestant ethic" insofar as both of them are a constellation of subjectively held beliefs and value judgments on the one hand and an objectively observable pattern of action on the other. The difference is that in describing the effect, Weber focuses more on the objectively observable pattern of action, and in describing the cause, he focuses on the subjectively held beliefs and value judgments. For Weber a particular configuration of beliefs and values (the "Protestant ethic) was the cause of the emergence of a particular set of habits (the "'spirit' of capitalism").

Just as he has a very particular understanding of the "individual," Weber has a very particular understanding of the term "cause." He acknowledges that a variety of causes could be identified (and probably would have to be identified) in order to explain the survival and eventual triumph of the spirit of capitalism over competing *Lebensfühurungen*. For example an inquiry into the cause of the success and expansion of capitalism today would have to identify "economic selection" as the decisive factor: "Through a process of economic *selection*, the capitalism that today dominates economic life socializes and creates the economic functionaries that it needs, both owners of businesses and workers" (PESC, 19). But identifying "economic selection" as the cause of the survival and expansion of capitalism has nothing to do with the causal question that Weber is interested in—Weber is interested in the causes of the origin of the spirit of capitalism. He draws a sharp distinction between the cause of the origin of a phenomenon and the cause of its survival and (more importantly) its spread. However critical a role "selection" may have played in the survival and spread of capitalism, it played no role whatsoever in its origin:

> The limitations of the notion of "selection" as a means to explain historical phenomena can be grasped here vividly. In order for a particular type of organized life and a particular conception of vocational calling adapted to the uniqueness of modern capitalism to be "selected" obviously they must first have originated among—and as a mode of thinking be carried by—*groups* of persons rather than simply isolated individuals. Hence it is the origin of this mode of thinking and its carrier groups, that actually needs to be explained. (PESC, 19)

Before the possibility of the survival and spread of capitalism, capitalism has to come into existence. A causal account of the existential origins of capitalism must dispense with the notion of "economic selection." In order for the notion of "selection" to make sense, "the origin of this mode of thinking and it carriers" need to be explained and "selection" makes no contribution whatsoever in this regard. Weber's goal is to put forth a testable hypothesis about what he thinks is quite significant—the historical origins of the unique individual that he has called the "'spirit' of capitalism."[2] Even more specifically, Weber is not interested in the origin of the "'spirit' of capitalism" as a theoretical idea in the minds of "a few isolated individuals" (i.e., some academician writing a book about the "theory of capitalism"). He is interested in the "'spirit' of capitalism" as a description of the observable pattern of action adopted by groups of people in the world, which in its turn became a cause for changing the world in which these people lived. Just as the ideal-type of the effect has to capture a unique individual and a particular pattern of action, the ideal-type of the cause has to capture a unique individual and a pattern of thought and beliefs.

The cause of the "'spirit' of capitalism" is a constellation of values and beliefs that come under the heading "Protestant ethic." In the most general terms we can describe the "Protestant ethic" as "the pursuit of the Kingdom of God." Weber is aware of the fact that he cannot end his inquiry by identifying "the pursuit of the Kingdom of God" as the causal origin of the "'spirit' of capitalism." This is due to the fact that besides the Protestants other groups of people in history have also identified the "pursuit of the Kingdom of God" as the highest value and shaped their empirically observable conduct according to this value—but they have not adopted the spirit of capitalism as a *Lebensführung*. Leaving aside the multitude of examples from outside of Christianity, we find prominent examples from within Christianity (and from within Protestant Christianity at that). The Hutterites and Amish are examples of Protestants who have pursued "the Kingdom of God" as the highest value but have not developed a *Lebensführung* that remotely resembles the "'spirit' of capitalism." Consequently, it is not sufficient to merely identify the "'spirit' of capitalism" as an effect of the "Protestant ethic"—those unique features of the "Protestant ethic" that made it the causal origin of the spirit of capitalism have to be identified.

Weber does this by looking at the way Luther interpreted the term "*Beruf*" and the way Calvin interpreted the doctrine of predestination. In the first chapter we saw the different obstacles that hindered the emergence of the "'spirit' of capitalism." Economic traditionalism and traditional religious piety put inner restraints on the pursuit of economic profit. Adventure capitalism broke all inner restraints on the pursuit, use, and display of economic wealth. The interpretation of Christian scriptures by Luther and Calvin played a critical role in breaking the hold of economic and religious traditionalism among the early Protestants on the one hand and tempering the ostentatious and irrational display of worldly wealth on the other. Consequently, it becomes clear that Weber is using the phrase "Protestant ethic" in a very specific sense—and it is only in its specificity that the "Protestant ethic" has any relationship to the "'spirit' of capitalism." The term "Protestant" in the phrase does not refer to the collectivity of denominations that emerged after 1517 and came to be known as the "Protestant Church" which consciously distinguished itself from the Catholic Church. It refers to specific seventeenth-century sects within the Protestant Church that adopted particular interpretations of the sixteenth-century teachings of Luther and Calvin. The term "ethic" in the phrase does not refer to the generic "pursuit of the Kingdom of God" by living a godly life but to the particular beliefs entailed in the seventeenth-century interpretations of the teachings of Luther and Calvin. Weber identifies the Westminster Confession (1647) and the sermons of Richard Baxter as examples of these historically significant interpretations.

As noted above, just as Franklin's print shop is an example of what Weber means by "'spirit' of capitalism," Baxter's sermons are an example of what Weber means by "Protestant ethic." Weber devotes special attention to the way that Baxter reinterpreted Calvin's understanding of the doctrine of predestination. The doctrine attributed absolute power, knowledge, and goodness to a transcendent God and utter impotence, ignorance, and depravity to mortal human beings. The distinction between the two was so profound that it was a complete mystery why any of these lowly mortals would be worthy of God's favor and grace. But it was indeed a fact that God had selected a few of these mortals to be among the elect and condemned the rest to eternal damnation—the reason why the few were chosen as the elect remains beyond the ken of human understanding. As originally formulated, this doctrine entailed grave fatalism, loneliness, and anxiety on the part of believers. But the intellectual and spiritual burdens inherent in the doctrine were originally mitigated by the exuberant enthusiasm that is characteristically found among the first few generations of new religious sects. The burden was further obscured by the fact that there was only a vague awareness of the logical consequences of the doctrine on the part of the early believers. With the passage of time, the logical consequences of the doctrine gradually became clearer: "As the original fervor and inspiration of the first generations of the faithful waned, these consequences of the Predestination decree became all the more unbearable. Baxter knew well that the harshness of this doctrine precluded its continued endorsement by most believers" (Kalberg, 1996, 59). Faced with the challenge of keeping the old faith alive under new historical circumstances, "Baxter undertook a revision that, Weber argues, launched the Protestant ethic" (Kalberg, 1996, 59). Baxter's revisions came in the form of establishing novel relationships among previously known creedal points and scriptural passages. These revisions sought to cultivate particular beliefs among the faithful that would be strong enough to seize their personal being in the same way that the first generations of believers had been seized by their beliefs. Kalberg's analysis shows that Baxter's revisions sought to cultivate three important beliefs among the believers. Firstly, since the world exists for the glorification of God, an abundance of health, wealth, and prosperity among human beings in the world contributes to the glorification of God because it manifests His grace and blessings in the world. Secondly, those human beings whose work in the world contributes to increase health, wealth, and prosperity among the human inhabitants "could view themselves as noble instruments—or tools—of His commandments and divine plans" (Kalberg, 1996, 59). Thirdly, since God would only choose the righteous elect for the noble task of manifesting His glory in the world, the following conclusion could be drawn: "Those believers capable of systematic work on behalf of God's plan could convince themselves that their strength to do so emanated from the favoring Hand of an omnipotent God—and the believer could conclude, God would favor only those he had chosen to be among the predestined [elect]" (Kalberg, 1996, 59). From this analysis it becomes clear that the term "Protestant ethic" does not refer to only a set of ethical values, it includes within it a particular set of sentimental beliefs. It is only by taking into account this set of beliefs that Weber can account for the fact that the "'spirit' of capitalism" emerged only among some Protestant sects and not among others—it emerged only among those Protestant sects that embraced these particular beliefs. In sum, the ideal-type "'spirit' of capitalism" is a pattern of habits, based on a set of values (the Protestant ethic), containing within it a set of beliefs (a particular interpretation of the teachings of Luther and Calvin).

At this point it is worth pausing and engaging in some second-order reflection. From the foregoing discussion we can surmise that the socio-cultural reality to which the ideal-type "'spirit' of capitalism" refers is composed of three elements: sentimental beliefs, ethical values, and observable actions. The fact that observable actions and subjective values are described using the ideal-type has been already noted. Before beginning the discussion on the doctrine of predestination and its historically significant interpretations, Weber notes: "We can only proceed by examining the religious ideas as ideal-types, namely, as constructed concepts endowed with a degree of consistency seldom found in actual history" (PESC, 55). In sum, the ideal-type is the conceptual tool that captures "religious ideas"—which are a constellation of relationships among beliefs, actions, and values. In addition to making it possible to conceptualize these three elements, Weber's reflections on the ideal-type allow us to identify the dynamic relationship between them. In this relationship, beliefs play a dual role. On the one hand, beliefs confer legitimacy on the values that serve as motives for action (manifested in a pattern of action). On the other hand, beliefs confer meaning and significance on the action itself. In other words the relationship of beliefs to action is as significant as the relationship of beliefs to values, but differently significant—beliefs confer meaning on action and legitimacy on values.

One may be tempted to claim that "beliefs" have entered the discussion because the subject under discussion is related to religion—the Protestant ethic. But a closer look at Weber's work shows that beliefs are an integral part of all social action that is based on the presupposition of the legitimacy of any social structure—whether that structure is related to religion or not. For example, Weber asks us to consider the "logical structure of the *concept of the state*" that is used in socio-cultural inquiry: "When we inquire as to what corresponds to the idea of the 'state' in empirical reality, we find an infinity of diffuse and discrete human actions, both active and passive, factually and legally regulated relationships, partly unique and partly recurrent in character" (OSS, 99). An investigator sees a group of human beings engaged in a bewildering variety of action to which they give the name "state." In an effort to come up with a scientific conception of the "state," the investigator begins with the presupposition that as bewildering as the variety of actions may be to an outside observer, in the minds of the actors these actions are "all bound together by an idea" (OSS, 99). Since there can be no scientific concept of the "state" without an adequate understanding of this idea, one has to look at the constituent elements that make up this idea. Weber notes that the "idea" is composed of "the belief in the actual or normative validity of rules and of the authority relationships of some human beings towards others" (OSS, 99). From this description we see that the concept of the "state" has to account for three elements:

- an observable pattern of action by a group of actors
- accepted rules and authority relationships that shape this pattern
- an affirmation of the validity of these rules and authority relationships.

The observable pattern of action is the social fact. The accepted rules and authority relationships are the values that serve as reasons for the fact for the actors. And the "affirmation of the validity of these rules and authority relationships" is the belief that legitimates the values and confers meaning on the facts. Weber's analysis shows us that wherever we can objectively observe patterned action in the socio-cultural domain, two different types of subjective elements are always and everywhere present, ethical value judgments and aesthetic/sentimental beliefs.

Another way to illustrate the presence and dual of belief in any type of habitual action is to look at the relationship between values and scientific inquiry. Weber notes that in one very specific sense, cultural values can be critically judged by science: "This criticism can of course have only a dialectical character, that is, it can be no more than a formal logical judgment of historically given value-judgments and ideas, a testing of the ideals according to the postulate of the internal *consistency* of the desired end" (OSS, 54). All social actors live within a social and natural environment. Their action in this environment is made possible by the acceptance of certain values. Scientific analysis can critically judge values in the sense that it can determine the internal consistency of pursuing a particular end motivated by a particular value in a given social-natural environment. The fact that science can judge the "internal consistency of the desired end" allows the actor to attain a type of understanding about his value commitment that he cannot get from anywhere else: "It can, insofar as it sets itself this goal, aid the acting willing person in attaining self-clarification concerning the final axioms from which his desired ends are derived. It can assist him in becoming aware of the ultimate standards of value which he does not make explicit to himself, or which he must presuppose in order to be logical" (OSS, 54).

When an individual consciously looks at the "internal consistency" of her values she will inevitably come face to face with the certain "final axioms" from which these values are derived. Very rarely are these "final axioms" explicitly stated or consciously understood, but they are everywhere and always the "ultimate standards" that the actor "must presuppose in order to be logical." Scientific analysis can help the actor attain self-clarification about the "final axioms" and "ultimate standards" on which her value commitments are based—but beyond this, scientific inquiry cannot go: "As to whether the person expressing these value-judgments *should* adhere to these ultimate standards is his personal affair; it involves will and conscience, not empirical knowledge" (OSS, 54).

Weber notes that observation of human action shows that human beings are constantly making value judgments. Decisions are made to pursue certain goals and forgo other goals in the face of the limitations and opportunities afforded by the natural and social environments. Logically speaking these decisions to "act thus when one could have acted otherwise" are not possible unless one presupposes some valid and legitimate standard that can be used to choose from among the different options. It is obvious that the choice itself cannot validate the standard (the value) that was used to make the choice—the validation of the value that made the value judgment possible comes from "final axioms" and "ultimate standards." When this point is expressed using nontechnical language, the relationship between belief and values becomes clear: "Only on the assumption of belief in the validity of values is the attempt to espouse value-judgments meaningful. However, to *judge* the *validity* of such values is a matter of *faith*" (OSS, 55). In other words, to affirm the validity of a value and express it in the form of a value judgment that judges some "phenomenon of human existence" as "good" or "bad" is ultimately a matter of faith. The sense in which science cannot judge values is precisely at this point—the point at which the normative validity of values is in question. Science can judge values as long as there are questions regarding their logical consistency but science must become silent when it comes to determining their normative validity (i.e., passing judgment on "good" and "bad")—this can only be done on the basis of faith and belief. Just as science cannot determine the normative validity of values, it cannot determine the cultural significance of social facts. Once again, as long as the question of *Sinn* (meaning) is limited to formal, logical consistency, science can pass judgment on *Sinn*—but once one enters the domain of cultural affirmation, that affirmation can only be conferred by belief.

The foregoing discussion lays bare the crucial role that beliefs play in empirical reality. Irrespective of whether we are talking about the economic, political, aesthetic, erotic, religious, or intellectual spheres of culture, belief is everywhere and always present. The pattern of action exhibited by the actors in each of these spheres is made possible by the affirmation of some (and the rejection of other) values. Since the affirmation and rejection of values is always and everywhere a matter of faith, belief is always and everywhere present in cultural activity. Weber's insights on the role of faith and belief in culture allow us to bring greater precision to some of the points that have been made above. Much of the discussion has focused on the fact that the ideal-type is a type of concept that makes it possible to capture beliefs, actions, and values. As the discussion progressed, it gradually became clear that in addition to capturing these elements of empirical reality, the ideal-type also makes it possible to identify the dynamic relationship between them. To the degree that the ideal-type has to capture the relationship between the three, it has to be able to capture belief (perhaps not as a "constituent" of empirical reality but the presupposition that makes empirical reality possible).

On the topic of concept formation, Weber's position is clear—the soft sciences employ concepts in the pursuit of scientific knowledge no less than the hard sciences. But given the ultimate ends that they are aiming for, the concepts used by the soft sciences are qualitatively different from those used by the hard sciences. As was the case with causality, on the issue of concept formation Weber was surrounded by contemporaries who took one of the two dominant positions on the issue. On the one hand, there were the positivists who took the position that the soft sciences should adopt the conceptual tools of the hard sciences. For this position there was no need for the soft sciences to use a different type of concept than the type used by the hard sciences. On the other hand, the intuitionists took the position that the soft sciences could dispense with concepts altogether. As was the case with the issue of causality, Weber rejected the reductionism of the positivists and the dualism of the intuitionist and put forward a relational alternative. The evidence and arguments that Weber put forth in support of his position provide greater insight into Weber's understanding of concept formation.

THE HISTORICAL SCHOOL AND THE LIMITATIONS OF "SYSTEM OF LAWS"

The position of the Historical School was that "the ideal which all the sciences, including the cultural sciences, serve and towards which they should strive even in the remote future is a system of propositions from which reality can be 'deduced'" (OSS, 73). In simpler terms the ultimate goal of scientific inquiry is to "discover" the "laws of nature" that (allegedly) govern empirical reality and then explain all observed events in empirical reality in terms of these laws. In taking this position the economists from the Historical School found support from certain voices in the natural sciences. Weber cites the example of "a leading natural scientist [who] believed that he could designate the (factually unattainable) ideal goal of such a treatment of cultural reality as a sort of 'astronomical' knowledge" (OSS, 73). For this position, the goal of the soft sciences should be to eventually produce a "system of laws" from which all of empirical reality could be "deduced" in the same way as astronomy had (allegedly) succeeded in producing a conceptual system from which all valid "astronomical knowledge" could be deduced. Weber says that we should not "spare ourselves the trouble of examining these matters more closely" (OSS, 73). He asks us to assume that the ideal that the positivists are aspiring to is actually attained: "Let us assume that we have succeeded by means of psychology or otherwise in analyzing all the observed and imaginable relationships of social phenomena into their ultimate elementary 'factors,' that we have made an exhaustive analysis and classification of them and then formulated rigorously exact laws covering their behavior" (OSS, 75). Having produced a "system of laws" in the soft sciences that is (allegedly) the same as that produced by "astronomical knowledge," the question that naturally emerges is: "What would be the significance of these results for our knowledge of the *historically* given culture or any individual phase thereof, such as capitalism, in its development and cultural significance?" (OSS, 75).

Weber details this point by carefully analyzing the relationship between the phrases "astronomical knowledge" and "system of laws." The question is whether "astronomical knowledge" can be precisely formulated in a "system of laws." For Weber there is a problem at the very beginning: "The first thing that impresses one is that the 'astronomical' knowledge which was referred to is not a system of laws at all. On the contrary, the laws which it presupposes have been taken from other disciplines like mechanics" (OSS, 73). There is no such thing as "astronomical" knowledge without presupposing the existence and validity of non-astronomical laws of mechanics. And for their part there is no such thing as the "laws of mechanics" without presupposing the existence and validity of the nonmechanical "laws of mathematics." The correspondence between "astronomical knowledge" and "system of laws" breaks down at the very initial step because in order to move from one to the other a presupposition has to be affirmed (the validity of mechanics and mathematics) that cannot be done either by "astronomy" or by "laws."

There is one other important disjuncture between "system of laws" and "astronomical knowledge." To the degree that astronomical phenomena capture our attention because of their uniqueness, to the same degree any given "system of laws" becomes irrelevant to the type of knowledge that is "*significant* for us" (OSS, 73). The particular constellation that is significant for us can only be explained with reference to another particular constellation—never with reference to a "system of laws": "Every individual constellation which [astronomical knowledge] 'explains' or predicts is causally explicable only as the consequence of another equally individual constellation which has preceded it. As far back as we may go into the grey mist of the far off past, the reality to which the laws apply always remains equally *individual*, equally *undeducible* from laws" (OSS, 73). Whether we are talking about astronomical phenomena or chemical phenomena, it is equally the case that as soon as the question of the phenomena's individuality and uniqueness emerges, any reference to "laws" and "systems of laws" becomes superfluous. It is almost self-evident that only on the basis of the presupposition that there is something unique about "chemical phenomena" that sets them apart from "astronomical phenomena" that sets both of them apart from "biological phenomena" that we have the sciences of chemistry, astronomy, and biology. The uniqueness of "chemical phenomena" and the individuality of chemistry cannot be derived from any "laws" or "system of laws." The conclusion is as clear as it is stark—even in the domain of physical-natural phenomena, "laws" and "system of laws" cannot produce a certain type of knowledge (knowledge of the unique individual that is significant from a particular viewpoint).

The theoretical goal of reducing or deducing all empirical reality to a "system of laws" is dangerous for at least one other reason. Just as a "system of laws" eliminates the unique individual from empirical reality, it also eliminates *Sinn* from empirical reality. By this point in the discussion it is obvious that interpretive understanding of *Sinn* is indispensible for reaching the ultimate goal of the soft sciences, that is, causal interpretation. Weber repeatedly and emphatically makes the point that the types of concepts employed by the hard sciences are completely inadequate means for attaining this end: "The *significance* of a configuration of cultural phenomena and the basis of this significance cannot however be derived and rendered intelligible by a system of analytical laws (*Gesetzesbegriffen*), however perfect it may be, since the significance of cultural events presupposes a *value-orientation* towards these events. The concept of culture is a *value-concept*" (OSS, 76). More considered reflection reveals that there is an intimate relationship between meaningfulness on the one hand and uniqueness and values on the other: "Meaningfulness naturally does not coincide with laws as such, the more general the law the less the coincidence. For the specific meaning which a phenomenon has for us is naturally *not* to be found in those relationships which it shares with many other phenomena" (OSS, 76f.). The reason that a phenomenon is culturally significant is because of its individuality and unique characteristics. The following observation offered by Weber identifies the source. Speaking of the attempt to resolve a social problem, Weber notes: "The broader its cultural *significance*, the less subject it is to a single unambiguous answer on the basis of the data of empirical sciences and the greater the role played by value-ideas (*Wertideen*) and the ultimate and highest personal axioms of belief" (OSS, 56). Since this point is self-evident, it would be worth pausing and illustrating it with an example. It is absurd to claim that there is some "law of religion" (or "law of human nature") that leads to the necessary conclusion that Buddhism is "objectively" superior to Christianity or Taoism. Similarly there is no "law of economics" (or "law of human nature") that demonstrates the "objective" superiority of capitalism over socialism or communism. Consequently, it is demonstrably absurd to claim that an actor has conferred *Sinn* (meaning) on Buddhism or capitalism because of some "law." Everywhere and in all cases, meaning is conferred upon "certain phenomena of human existence" by "value-ideas (*Wertideen*) and the ultimate and highest personal axioms of belief."[3] Since it is logically absurd to claim that actors confer meaning on a particular cultural phenomenon because of some "law," it is equally absurd to claim that the investigator can put forward a scientific account of the phenomenon exclusively in terms of "law." Because the act of meaning bestowal by the actor is always from a particular "evaluative standpoint" (i.e., commitment to certain values on the basis of certain beliefs), any scientific account of cultural phenomena has to understand, explain, and (causally) interpret this standpoint.

Weber notes that certain developments in the sciences (some going back to classical Greece and others as recent as the nineteenth century) have collectively created a situation where the "evaluative standpoint" has been progressively eliminated from scientific inquiry (OSS, 85). Speaking of the more recent developments, Weber notes: "As rational analysis of society arose in close connection with the modern development of natural science, so it remained related to it in its whole method of approach" (OSS, 85). Looking at the method of inquiry of natural science that the emergent soft sciences sought to mimic, we find that this method is based on a "hope" that can be traced back to ancient Greece:

> In the natural sciences, the practical evaluative attitude toward what was immediately and technically useful was closely associated from the very first with the hope, taken over as a heritage of antiquity and further elaborated, of obtaining a purely "objective" (i.e., independent of all individual contingen-

> cies) monistic knowledge of the totality of reality in a *conceptual* system of metaphysical *validity* and mathematical *form*. (OSS, 85)

Weber uses the phrase "will-to-believe of naturalistic monism" (OSS, 86) to describe the hope that "the totality of reality" could be captured in a "conceptual system of metaphysical *validity* and mathematical *form*." He identifies modern biology's conversion to the doctrine of "naturalistic monism" as a watershed event in the history of modern science. After this event the "evaluative standpoint" in scientific inquiry would die a natural death:

> When modern biology subsumed those aspects of reality which interest us *historically*, that is, in all their concreteness, under a universally valid evolutionary principle, which at least had the appearance—but not the actuality—of embracing everything essential about the subject in a scheme of universally valid laws, this seemed to be the final twilight of all evaluative standpoints in all the sciences. (OSS, 86)

After biology's embrace of naturalistic monism "it appeared as if there was in general no conceivable meaning of scientific work other than the discovery of the *laws* of events" (OSS, 86). Under these circumstances the "scientifically essential" came to have the meaning that is often associated with it: "Only those aspects of phenomena which were involved in 'laws' could be essential from the scientific point of view and concrete 'individual' events could be considered only as 'types' (i.e., as representative illustrations of laws. An interest in such events in themselves did not seem to be a 'scientific' interest") (OSS, 86). In other words only the generic, recurrent, and universal aspects of empirical phenomena "could be essential from the scientific point of view" and the concrete, unique, and individual aspects of empirical phenomena were deemed scientifically inessential. At the level of "theory and theoretical conceptualization" (OSS, 85), "naturalistic monism" remained largely neutral with respect to meaning and values. But as the foregoing discussion makes obvious, it took a definitive stand on the concrete individual. From this monistic (or reductionist) perspective, the concrete individual is of scientific significance only to the degree that it is an illustration of a universal "law" or the example of generic "type."

The advantages that the "will-to-believe of naturalistic monism" conferred on the natural sciences are well known—a significant part of the remarkable progress in the hard sciences is a product of this very belief. The disadvantages are not as well known. Weber's analysis brings us face to face with these disadvantages. One of the disadvantages has been hinted at. By mimicking the method of natural science and setting a "system of laws" as the end of scientific inquiry, the emergent soft sciences precluded the possibility of appreciating the significance of qualities, sentiments, values, individuals, variation, and significance in empirical reality. Let us imagine that the soft sciences came up with a method of scientific inquiry that precluded the recognition of atoms, cells, DNA, and galaxies from being even recognized. And if the method did allow such phenomena to be recognized, it contained an a priori condition that these physical-natural phenomena be understood only as functions of beliefs, values, qualities, and motives. How "objective" and "scientific" would such a method be? This is precisely the situation that emerges if the conceptual tools of the hard sciences are used to study socio-cultural phenomena. The logical and intellectual absurdity here is self-evident—and for this very reason requires more clarification, but will have to wait until the next chapter.

As debilitating as the shortcoming of trying to describe socio-cultural phenomena using the conceptual tools borrowed from the hard sciences is, it may not be the most debilitating disadvantage of the "will-to-believe of naturalistic monism." A more debilitating disadvantage of this naturalistic prejudice is described by Weber in these words: "The nature of that *Weltanschauung* with its optimistic faith in the theoretical and practical rationalizability of reality had an important consequence insofar as it *obstructed* the discovery of the *problematic* character of that standpoint which had been assumed as self-evident" (OSS, 85). Let us (along with the legions of hard scientists and their imitators in the soft sciences) take a blind leap of faith and assume that the totality of reality can be captured by a conceptual system of mathematical form. It is at this point that the deeply problematic element in this standpoint comes to the fore—what is the basis of the "hope" that animates the very quest for scientific knowledge? This "hope," like any other cultural value, cannot be derived from any "system of laws" that science has discovered (or will ever discover) nor is it based on anything that is inherent in human nature. Given this fact, the question naturally emerges: What is this hope based on or what is the basis on which the value of scientific knowledge is to be affirmed? The *Weltanschauung* emerging from the "will-to-believe of naturalistic monism" precludes the possibility of even asking this question. Consequently, the most egregious consequence of a theoretical position that eventually leads to "twilight of all evaluative standpoints" in scientific inquiry is that this position becomes blind to the values (and the accompanying beliefs that justify these values) that it itself is based upon.

In sum, Weber critiques the attempt of the positivists to reduce all socio-cultural reality to a "system of laws," along the lines of the hard sciences, for a number of reasons. The most basic is logical. Given the very nature of empirical reality, the infinitely smallest segment of this reality cannot be subsumed under any single law or any system of laws. The second important reason is methodological. Reducing reality to a system of laws eliminates the most important features of empirical reality from the domain of scientific inquiry: qualities, sentiments, and values that collectively make up the unique individual. The third reason brings the logical and methodological into relationship with each other—the fact that scientific knowledge is a product of human culture and not of any laws of nature or anything inherent in human nature remains hidden. Consequently, it is practically impossible to conduct scientific inquiry into the qualities, sentiments, and values that collectively make up the unique individual—keeping in mind that science is a unique way of attaining knowledge both in terms of its means (concepts) and ends (causal interpretation). As a result, scientific inquiry continues to be carried on the basis of vague, dimly understood, and naïvely affirmed value commitments and beliefs. Having looked at Weber's analysis and critique of the positivists on the topic of concept formation, we turn to his analysis and critique of the intuitionists.

PEIRCE'S PHENOMENOLOGY AND WEBER'S CRITIQUE OF INTUITIONISM

On the issue of concept formation, Benedetto Croce can be seen as a representative of the intuitionist position. The intuitionists claimed that socio-cultural inquiry could dispense with the use of concepts altogether. Croce claims that knowledge of individual, concrete objects is only possible through an intuitive grasp of the object—or that knowledge of an object can be only intuited. He argues that while objects are intuitions, relationships between objects are concepts. The concept can have only "a general and therefore abstract nature" (RK, 167). Because the concept is abstract it follows that individual, concrete objects cannot "be subsumed under concepts. They can only be 'intuited.' Knowledge of objects, therefore, is only 'artistically' possible" (RK, 167). This understanding of concept makes abstract concepts less than empirical reality in the sense that no particular object can ever be subsumed under any concept—the object will be always "more" than the concept. Weber notes that, given Croce's definition of "concept," "A 'concept' of the concrete is a contradiction in terms" (RK, 167). The implications of this understanding of the concrete are far-reaching for those soft sciences, like history, that study unique individuals. Weber notes that if it is indeed the case that knowledge of concrete objects can be obtained only through "intuitions," then it naturally follows that the validity of any knowledge claim "cannot be resolved by any discursive analysis" (RK, 167). This is due to the fact that socio-cultural knowledge "cannot be the object of a 'logical' critique, for logic concerns only (general) concepts and their definitions" (RK, 167).Consequently, the implications of the intuitionist position go beyond particular knowledge claims about particular objects. If this position is accepted, then socio-cultural inquiry cannot be subjected to analysis and critique using the tools of logic. For Weber this means that scientific inquiry into socio-cultural phenomena is not possible. Weber demonstrates that the soft sciences do employ concepts (even though they are different than the ones employed by the hard sciences). Furthermore, a logical analysis of the validity of the conclusions reached by using these concepts is indeed possible, as is the adequacy of the concepts used to carry out the inquiry.

Weber begins his critique of Croce by noting that his views are the result of a psychologism rooted in a variety of naturalistic errors. Firstly, relational concepts are composed of "exactly as much 'intuition' of the immediate experience of everyday life as any concept of an object" (RK, 168). Furthermore, even an "exact" natural science like physics employs concepts that challenge Croce's definition of concept as being only relational. Secondly, the converse of this claim, that "concepts of objects" are not really concepts but "intuitions," is the result of conflating two different meanings of "intuitability." Weber notes: "The intuitive self-evidence of mathematical propositions is quite different from the 'intuitability' of the multiplicities of 'experience' immediately given 'in' us and 'external to' us, experienced and accessible to experience. To use Husserl's terminology the distinction is between 'categorical' intuition and 'perceptual' intuition" (RK, 168).

Weber goes on to note any claim that knowledge of empirical reality is solely the product of intuitions or a direct copy of empirical reality is based on (quite elementary) logical errors. He notes that the "experience" of first-person events cannot be grasped by this type of intuition: "As soon as it becomes an object of *thought*, not even a first-person experience can be 'reflected' or 'reproduced' in inner experience. That would not constitute thought *about* the experience, but rather another 'experience' of the earlier experience. Or rather, since this is impossible, a *new* 'experience,' 'accompanied' by the 'feeling' . . . that one has already 'experienced' 'this' . . . before" (RK, 169). In this short but exceedingly dense and important passage, Weber identifies three distinct types of "experience." The defining characteristic of the first type of experience is that it is "that which is immediately given to 'experience' and which remains indeterminate" (RK, 169). Second, there is "experience" as ordinarily understood—any phenomenon that gives birth to the interrogative question "What is that?" or the declarative proposition "That is/was ____." Third, "experience" is the thought, idea, etc., that is generated as a result of the question or proposition. The answer or statement that is "experience" in the third sense is invariably and irreducibly "conceptual" in character. In other words there is a type of experience that is purely conceptual—and for Weber it loses none of its experiential character. Referring to propositions about empirical reality that aspire to "validity," Weber notes that logical operations are involved in all such propositions: "Although these logical operations do not include the 'justification' of general concepts, they certainly do include their constant *use*. Therefore they include abstraction and comparison" (RK, 169). We can derive the following typology of "experience" from the above discussion:

1. Experience as such.
2. Experience as experiencing a brute fact.
3. Experience as a thought about experience as such or experiencing a brute fact.

Expanding on Huff's observation that there is remarkable similarity between Weber's notion of causality and Peirce's notion of abduction, we can easily see that there is remarkable similarity between the typology of experience that is implicit in Weber's critique of the positivists and intuitionists and Peirce's phenomenology. Peirce uses the terms Firstness, Secondness, and Thirdness to describe the three categories of his "particular phenomenology." Peirce's explication of the three categories leaves little room for doubt that they are a detailed, technical description of what Weber has called (respectively) experience as such, experience as brute fact, and experience as thought. Summarily stated, this is Peirce's description of Firstness, Secondness, and Thirdness:

- Category the First is the Idea of which is such as it is regardless of anything else. That is to say, it is a *Quality* of Feeling.
- Category the Second is the Idea of that which is such as it is as being Second to some First, regardless of anything else and in particular regardless of any Law, although it may conform to a law. That is to say, it is *Reaction* as an element of Phenomenon.
- Category the Third is the Idea of that which is such as it is as being a Third, or Medium, between a Second and its First. That is to say, it is *Representation* as an element of the Phenomenon. (5.66)

We will look at the detailed description of each of the three categories in turn.

The Category of Firstness

The category of Firstness refers to the *presentness* of the idea. This *presentness* makes itself felt as a quality of feeling. This feeling of *presentness* is not an abstraction, it is a "feeling" that excludes everything and all else. This feeling is independent of time, space, memory, vividness, etc.—in short, "it is such as it is quite regardless of anything else" (5.44). This feeling is also prior to any analysis being applied to it, whether that analysis is in the form of rationalization, categorization, judgment, etc. Peirce notes: "The unanalyzed total impression made by any manifold not thought of as actual fact, but simply as a quality as simple positive possibility of appearance is an idea of Firstness . . . The idea of the present instant, which, whether it exists or not, is naturally thought as a point of time in which no thought can take place or any detail be separated, is an idea of Firstness" (8.329). The following observation shows that Firstness does not refer to an "event" but a "state": "A feeling, then, is not an event, a happening, a coming to pass, since a coming to pass cannot be such unless there was a time when it had not come to pass; and so it is not in itself all that it is, but is relative to a previous state. A feeling is a *state*, which is in its entirety in every moment of time as long as it endures" (1.307). Other terms that describe Firstness can be gleaned from Peirce's detailed description of feeling. He notes that "feeling is nothing but a quality and a quality is not conscious, it is a mere possibility" (1.310). In sum, Firstness is a "state," "feeling," "quality," "possibility," independent of any descriptor of any type being attached to it.

Peirce's description of Firstness brings clarity to Weber's observation that there is a type of experience that is not experience in the sense of "having" an experience but "the experience which we simply *are*" (RK, 163). Weber explicitly describes this type of experience as experience "in the prescientific sense" (RK, 261, n.61). In Peirce's terms, experience in the "pre-scientific sense" is experience as Firstness because we are speaking of experience, per se, without any qualifier or descriptor of any type. Weber notes that prescientific experience "is constituted by the totality of our 'perceptions' in connection with undifferentiated 'feelings' and 'desires' that are associated with them. Put another way, the 'object of immediate experience' is constituted by the 'commitments' which we make at every moment and of which we are, at any given moment, 'conscious' in a very different degree and sense" (RK, 160). Experience in the first sense "cannot be the object of a proposition—that is, an empirical explanation of facts" (RK, 160).[4]

The Category of Secondness

Peirce and Weber use virtually the same words in describing the second type of experience, or Secondness: "brute fact." Both of them also explicitly state that these brute facts are facts of "internal" experience as well as "external" experience. As Weber notes, it is not the internality or externality of the experience that is important here, it is the fact that the "brute 'fact'" upsets our expectations and makes us aware of something other than ourselves. Going along with the intuitionists, Weber proposes that we distinguish between "immediate experience" and "observational experience" in the following manner: "Suppose that 'the object of immediate experience' is to be understood as the 'psychical' event 'in' us, as opposed to the totality of events 'external' to us—regardless of how the distinction between 'internal' and 'external' is drawn. And suppose that this 'psychical' event is to be understood as an object of valid empirical knowledge" (RK, 160f.). Irrespective of how we draw the distinction between "internal" and "external," for

Weber there is no logical difference between "immediate experience" and "observational experience." Weber notes:

> Every putatively valid piece of knowledge concerning the concrete complexes of immediate experience rests on "observation" of exactly the same logical structure as the "observation" employed in any analysis of the "objectified" world. In the first place, human conduct, the object of interpretation, invariably includes components which must simply be taken as brute "facts," just like the corresponding properties of any "object." (RK, 161)

Just as the "objective" world of nature contains brute facts, so does the "subjective" domain of human consciousness. Taken together, pre-scientific experience and experience as brute fact are examples of "immediate experience." Though they share the common trait of immediacy, they are immediate in two very different senses. The immediacy of the pre-scientific experience establishes the reality of experience as such. The immediacy of experience as brute fact suggests that there is something more than just experiencing consciousness in the universe. Experience as brute fact opens up the possibility that there are other consciousnesses in the universe besides the "subjective" one.

Secondness is the type of experience that makes us conscious of a non-self in the universe and it always involves two, struggle, surprise, and disruption. He describes the category of Secondness in these terms: "The second category that I find, the next simplest feature common to all that comes before the mind, is the element of struggle. . . . By struggle I must explain that I mean mutual action between two things regardless of any sort of third or medium and in particular regardless of any law of action" (1.322). Secondness refers to brute action in which one thing merely acts upon another "regardless of any law of action" because "so far as the idea of any *law* or *reason* comes in, Thirdness comes in" (8.330). Peirce elaborates: "When a stone falls to the ground, the law of gravitation does not act to make it fall. The law of gravitation is the judge upon the bench who may pronounce the law till doomsday, but unless the strong arm of the law, the brutal sheriff, gives effect to the law, it amounts to nothing. . . . The stone's actually falling is purely the affair of the stone and the earth at the time" (8.330). When two things exist in and of themselves, independent of each other, they do so without any effort and without facing any resistance or opposition. But the moment the two come into contact with each other, Peirce posits that shock and surprise are the results, and initially this surprise often gives rise to resistance on the part of the two things. There can be no effort that is not accompanied by a reaction (or resistance). Conversely, there can be no resistance/reaction without effort. Peirce notes that action/resistance or effort/resistance "are only two ways of describing the same experience" (1.324). This interplay of effort/action and resistance/reaction provides evidence of a double consciousness that makes us aware of an "inner" and "outer" world. This double consciousness accustoms us to think that there is an internal world of which we are the masters, which is the world of "our" inner consciousness. But we are also aware of an "other" external world over which we have no control. We see the internal world as "I" or "ego," and the external world as the "other." But by coming into contact with the "other" (the natural world) we become aware of the fact that the "I" or the "ego" itself is composed of a "self" and "other." Peirce notes:

> We become aware of ourselves in becoming aware of the not-self. The waking state is a consciousness of reaction; and as the consciousness itself is two-sided, so it has also two varieties; namely, action, where our modification of other things is more prominent than their reaction on us and perception, where their effect on us is overwhelmingly greater than our effect on them. . . . To this element I give the name Secondness. (1.324)

While this double consciousness is disturbing and surprising it is also "the pivot of thought" and the most effective of teachers. It is because of the dual nature of consciousness that it becomes possible for "the world of fact or *experience*" (1.321) to become a modifier of our internal world. When Peirce refers to a scientific intelligence being an intelligence that is "capable of learning by experience" (2.227), he is primarily referring to experience as Secondness. As long as consciousness does not encounter any surprises in the world of fact, it is content with itself, its assumptions, and its worldview. In other words in the absence of Secondness the self remains in a naïve relationship with itself and the world around it. It is only when an experience produces results that contradict the consciousness's expectations that its sense of complacency is shaken. Its instinctive reaction is to attribute the surprise/contradiction to the outer world (the world of nature). On closer scrutiny we find that the surprise is not attributable to external causes but rather to the contradiction of internal expectations. Peirce describes the dynamics of *surprise* in these words:

> A man is more or less *expecting* one result and suddenly finds something in contrast to that forcing itself upon his recognition. A duality is thus forced upon him: on the one hand, his recognition which he had been attributing to Nature, but which he is now compelled to attribute to some mere inner world and on the other hand, a strong new phenomenon which shoves that expectation into the background and occupies its place. (5.57)

Surprise serves as a catalyst for consciousness to move beyond mere feeling, sentiments, and beliefs and in the direction of passing perceptual judgments on phenomena. For both Weber and Peirce, experience as Secondness is the first step and a necessary condition for scientific inquiry—but is far from being a sufficient condition. Scientific inquiry, in the proper sense, begins when we have experience in the third sense.

The Category of Thirdness

Just as there is experience beyond experience as such and experience as brute fact for Weber, for Peirce "it is impossible to resolve everything in our thoughts into those two elements [of Firstness and Secondness]" (1.343). In addition to feelings and brute experience, "we constantly predict what is to be. Now what is to be, according to our conception of it, can never become wholly past. In general, we may say that *meanings* are inexhaustible" (1.343). A category is needed to account for the ideas that are not merely feeling or experience in the present and past but allow us to make predictions about the future in light of the experiences of the present and the past. This is the category of Thirdness. Just as Firstness is what it is by virtue of its presentness and Secondness is what it is by virtue of encounter, Thirdness "is that which is what it is by virtue of imparting a quality to reactions in the future" (1.343).

The similarity between Weber's description of experience in the third sense and Peirce's category of Thirdness is deepened in light of the fact that this category can be applied to the "inner" world as well as the "outer" world. For Weber scientific knowledge of experience inside of us and outside of us, experience of human emotions, natural events, physical phenomena, psychical phenomena—in short, knowledge of experience in every sense of the word, without any qualification, is possible only after the experience has elapsed and become the object of rational reflection. Weber notes:

> The leaden diffuseness of "immediate experience" must be broken . . . in order to take even the first step toward a genuine understanding of [even] *ourselves*. The claim that every "experience" is perfectly certain obviously shows *that* we experience. However, *what* we really experience can become accessible to "interpretation" only after the "experience" itself has elapsed. In which case what is experienced can become the "object" of a proposition. (RK, 162)

Once the experience has elapsed, there is no way of reproducing it. Knowledge of it can be obtained only after the experience has become the "object" of a proposition, and once this happens, abstract concepts inevitably come into play. This is due to the fact that in a very particular sense the thought about the experience (in contrast to the experience as such) is a thoroughly synthetic construct because its "unity" is constituted by the selection of those aspects from immediate experience that are deemed "essential" from a particular point of view. (It is worth recalling Weber's discussion of separating the causally "essential" and "inessential" factors in a particular case.) There is no sense in which this synthetic construct can be considered a reproduction or a reexperiencing of the experience as such.

In Peirce's terms, Weber's "synthetic construct" is a "law" or "reason" that serves as the "medium or the connecting bond" (1.337) between two ideas, two experiences, two feelings. Law and reason also establish the bond between an idea and an experience, or an experience and a feeling, or a feeling and an idea. Peirce argues that law and reason can never be reduced to a dyadic relation because a dyadic relation can only account for a brute experiential fact, while law or reason introduces a mental element that seeks to offer an explanation or justification for the brute experiential fact. Any judgment regarding an experiential fact is composed of an element that is not given or present in the experience itself. Putting the issue in algebraic terms, Peirce notes:

> The criticism which I make on that algebra of dyadic relations, with which I am by no means in love, though I think it is a pretty thing, is that the very triadic relations which it does not recognize, it does itself employ. For every combination of relatives to make a new relative is a triadic relation irreducible to dyadic relations. Its *inadequacy* is shown in other ways, but in this way it is in a conflict with itself *if it be regarded . . . as sufficient for the expression of all relations*. (8.331)

Consequently, Thirdness has "the character of mediation, continuity and relationality" (Ochs, 1998, 109). It manifests itself in the form of law or reason that establishes a relationship between inner/outer, first/last, past/future and other such dichotomies. The following is a summary of the three categories:

> It seems, then, that the true categories of consciousness are: first, feeling, the consciousness which can be included with an instant of time, passive consciousness of quality, without recognition or analysis; second, consciousness of interruption into the field of consciousness, sense of resistance, of an external fact, or another something; third, synthetic consciousness, binding time together, sense of learning, thought. (1.377)

The category of Thirdness is an apt description of Weber's observation that in addition to the experience that we are and the experience of brute facts, there is the experience of making propositional statements. This latter type of experience cannot be confused with the former two. In a very specific sense this is the type of experience that scientific inquiry is primarily concerned with—it seeks to establish the validity of the propositional statements that we make about experience. As Weber has detailed, it is not possible to make such statements without the aid of "empirical rules" or "causal generalizations"—which are examples of Thirdness.

Table 3.1.

Weber's Typology of Experience	Peirce's Phenomenological Categories
Pre-scientific experience, the experience that "we are."	Firstness: Feeling, possibility, spontaneity, freedom.
Brute force.	Secondness: Struggle, reaction, force, competition, necessity.
Thought about pre-scientific experience or about experience as brute force.	Thirdness: Habit-law, generalizing tendency, cooperation.

In his engagement with the positivists and intuitionists on the issue of concept formation, Weber expends a significant amount of energy going into seemingly pedantic discussion of arcane, philosophical minutiae. But when we translate Weber's critique of the positivists and intuitionists into the terms of Peirce's phenomenology, we can see why this is far from a pedantic, arcane exercise. This translation is provided in the table above. Weber's analysis shows that (in Peirce's terms) the positivists conflate Secondness and Thirdness. This conflation has three important outcomes. Firstly, it allows causal agency to be attributed to the "laws of nature." Secondly, it puts the "laws of nature" beyond the domain of scientific scrutiny by making it even theoretically impossible to inquire into the reasons, causes, and meaning of these laws. Thirdly, it makes it difficult for the positivists to notice and accept the significance of Firstness in not just human experience and empirical reality but also in the process of scientific inquiry—scientific inquiry itself being a unique, particular segment of human experience and empirical reality. The intuitionists for their part conflate Firstness and Thirdness. This conflation also has three outcomes. Firstly, it makes it possible to make something in "human nature" the sole causal agent in the universe. At the same time, it places the "we are" that is carrying out the scientific inquiry beyond the scope of scientific scrutiny. Thirdly, it makes it marginalize the significance of Secondness in terms of its place in empirical reality and scientific inquiry.

The clarification of Weber's insights on the different types of experience by Peirce's phenomenology also allows us to see the nuanced stance that Weber has toward the positivists and intuitionists. He neither rejects their respective positions nor accepts them. With respect to the positivists, Weber does not challenge the claims that "laws" play an important role in scientific inquiry—for Weber, scientific inquiry, by definition, is impossible in the absence of something like "laws." He does reject the conflated positivist understanding of "laws" and the role that they accord them in scientific inquiry. Similarly, Weber does not reject the claim that "intuition," "imagination," "inspiration," or other peculiarly human traits play an important role in scientific inquiry. He challenges the intuitionist understanding of the role that these elements play. Weber's discussion of the different types of experience is important in this context because it will help us to understand the role and limitations of "objective" laws and "subjective" intuition in scientific inquiry (to be detailed in chapter 5).

This chapter began with detailing Weber's argument that the soft sciences employ a particular type of concept in scientific inquiry—the ideal-type. Weber has put forth the argument that the ideal-type is a genuine concept that is uniquely suited to study those aspects of empirical reality that are significant from the perspective of the soft sciences—that is, qualities, belief, values, and the unique individual. We have also seen that Weber critiques the position of the positivists that the soft sciences should adopt the conceptual tools of the hard sciences. For Weber, this aping of the hard sciences erases the very dimension of empirical reality that is significant from the perspective of the soft sciences. At the same time, Weber critiques the position of the intuitionists that we do not need to employ any conceptual tools in socio-cultural inquiry. For Weber, any inquiry into socio-cultural reality without the use of concepts is, by definition, not scientific. Moving beyond the descriptive analysis offered in this chapter, the next chapter will reconstruct Weber's position on concept formation.

NOTES

1. I have used Parson's translation "historical individual" in the place of Kalberg's "a specific *historical case*."

2. If the phrase "economic selection" is considered a modification of the phrase "natural selection," then it appears that Weber sees Darwin's account of the origin of new species to be woefully inadequate. The title of Darwin's book directly links "natural selection" with the "origin of species." The original title of the book was *The Origin of Species by Means of Natural Selection or the Preservation of Favored Races in the Struggle for Life*. Considering rational, modern capitalism a species of capitalism, Weber explicitly rules out "economic selection" as the "means" or "cause" of the origin. Peirce offers a theory of evolution that builds on the evolutionary theories of Darwin, Lamarck, and Spencer. It would be worth exploring if the Peircean alternative serves as a more adequate framework for Weber's account of the origin of capitalism. For a comparison and contrast of the Peircean and Darwinian theories of evolution, the reader may be interested in looking at a paper titled "A Framework for Evolutionary Algorithms Based on Charles Peirce's Evolutionary Semiotics" by Junaid Akhtar et al (2013).

3. The reader may be interested in noting the parallel between Weber's notion of *Wertideen* and the ideas of Max Kadushin (1895–1980) of the Jewish Theological Seminary. A student of the Talmud, who also studied Peirce and Whitehead, Kadushin offered a way of looking at society as value-infused (Rabbinic society in particular as centered around values). This echoes Weber's notion that the study of "social facts" is not possible without inquiry into values that these facts embody. Kadushin argued that "social values" are also "social facts" and vice versa; to study the one is to study the other. To study these values/facts, the inquirer must be guided by what he termed "value-concepts"—it seems that there is significant similarity between Kadushin's "value-concepts" and Weber's "value-ideas." Kadushin (1952), *The Rabbinic Mind*, New York: Bloch.

4. The fact that pre-scientific experience cannot be the object of a proposition does not in any way diminish its significance or reality. To belittle it or to disregard it is to do away with the very "we *are*" that is carrying out the inquiry. Even though it remains indeterminate in

scientific terms and irrelevant to empirical knowledge claims, pre-scientific experience is no less real than any other type of experience. It seems curious that Weber would go to such lengths to assert the reality of pre-scientific experience even though it is not directly relevant to empirical knowledge claims. As the details of Secondness and Thirdness will show, pre-scientific experience is the presupposition that makes scientific inquiry possible. This is the type of experience that makes it logically possible to ask the question "What is that?" or make the statement "This is x." In the absence of pre-scientific experience both the question "What is that?" and the statement "This is x."—thereby scientific inquiry itself—is rendered logically impossible.

Chapter Four

Weber on Concept Formation

A Reconstruction

Having explored Weber's conception of ideal-type by using an example, this chapter will explore the ideal-type from the perspective of second-order reflection. The first part of this chapter will turn to Weber's writings and make explicit the typology of concept that is implicit in his work. We will find that Weber's reflections on concept formation cannot be separated from his reflections on the different types of science and allow us to clearly see that Weber correlates the type of concept that a particular science employs with the type of knowledge that it aims for. The second part of the chapter will use Peirce's semiotics to thicken the description of Weber's typology of concepts. A look at the way Peirce describes the sign and the different types of signs (i.e., icon, index, symbol) shows that an ideal-type is a genuine symbol, whereas the concepts used by the hard sciences are examples of degenerate symbols. The chapter will conclude by noting that combined with Peirce's phenomenology in the previous chapter, Peirce's semiotics gives Weber the language to clearly describe not only the different types of experience and concepts but also the different ways of knowing them.[1] By the end of the chapter it will be clear that just as the soft sciences employ the category of causality "in its full meaning," they also employ the category of concepts "in its full meaning." In laying bare the degenerate and truncated understanding of concept in the hard sciences, Weber's insights contribute to the emergence of a more scientific conception of "concept."

WEBER'S TYPOLOGY OF THE SCIENCES AND CONCEPTS

For Weber *Sinn* (meaning) is more fundamental to socio-cultural reality than atoms are to physical-natural reality. It might be possible to give a definition of the reality that is studied by physics and not include the word "atom" in the definition—but one cannot give a definition of "culture" without including "meaning" in the definition. Weber defines "culture" as "a finite segment of the meaningless infinity of the world process, a segment on which *human beings* confer meaning and significance" (OSS, 81). Meaning is not only part and parcel of "culture" in general, it is always and everywhere part of any activity that a human being engages in (for example, the activity of scientific inquiry).[2] Weber notes that there is no such thing as scientific inquiry in the absence of an attitude that judges certain phenomena as being meaningful and significant and others to be meaningless and insignificant: "Whatever may be the content of this attitude—these phenomena have cultural significance for us and on this significance alone rests its scientific interest" (OSS, 81). Consequently, there is a twofold relationship between scientific inquiry and *Sinn*. On the one hand, scientific inquiry seeks to understand the *Sinn* that cultural beings confer on their actions—the actions that they variously label as "politics," "religion," "sport," "medicine," etc. On the other hand, each scientist chooses a particular object of inquiry (for example the cause for the rise of capitalism in the West) because he finds this topic to be significant for some reason. Since the meaning of the scientists' choice cannot be explained in terms of any "law of nature" or anything inherent in "human nature," science should be able to inquire into the reasons and meaning of the investigator's choice. In more general terms, science is (or more accurately should be) concerned with studying the meaning (*Sinn*) of science as a cultural activity.

Just as the very definition of empirical reality from the perspective of the soft sciences must contain the word "meaning," the very definition of "knowledge" must contain some word that refers to uniqueness and individuality in some way. Weber notes: "Knowledge of *cultural* events is inconceivable except on a basis of the *significance* which the concrete constellations of reality have for us in certain *individual* concrete situations" (OSS, 80). This point makes it clear that the terms "concrete" (as opposed to "abstract"), "particular" (as opposed to "universal"), and "finite (as opposed to "infinite") refer not only to what is known but also the context in which it is known—that which is known and the context in which it is known are both uniquely individual. Weber emphasizes, again and again, that uniqueness is that particular part of empirical reality that the soft sciences are interested in: "The type of social science in which we are interested is an *empirical science* of concrete *reality* (*Wirklichkeitswissenschaft*). Our aim is the understanding of the characteristic uniqueness of the reality in which we move" (OSS, 72).

The soft sciences aspire to study the significance of unique individuals in empirical reality from two different perspectives: "We wish to understand on the one hand the relationships and the cultural significance of individual events in their contemporary manifestations and on the other the causes of their being historically *so* and not *otherwise*" (OSS, 72). On the one hand, the inquiry seeks to study the characteristics and relationships of the unique individuals among themselves. On the other hand, it puts forth the investigative question: "Why are these characteristics and relationships as they are when they could be otherwise?" As was noted in the second chapter, the *Sinn* of economic profit has changed significantly in Western culture over the past five hundred years from necessary evil, to sign of salvation, to salvation itself. At the same time, the relationship between the unique individual known as "'spirit' of capitalism" and the unique individual known as "Protestant ethic" has also changed radically. The soft sciences must be able to study the causes of this variation in *Sinn*, characteristics, and relationships of the unique individuals that collectively constitute the empirical reality. Firstly, some of the most unique, distinguishing features of Western civilization are directly related to this cultural development. Secondly, his personal investigative interests are directly tied to understanding the causes of this variation because this variation has closed down the possibility of affirming certain values and opened up the possibility of affirming other values by the cultural beings who are living amidst the change. For Weber the phenomenon of "economic profit" is scientifically interesting precisely because of the variation of its *Sinn*. If it had been the case that always and everywhere the meaning of a particular cultural phenomenon was the same, then that phenomenon would be relatively uninteresting for Weber. The only thing that would make this particular cultural phenomenon even less interesting for Weber is if the variation in meaning was the result of some "law of nature" or something inherent in "human nature." The soft sciences go beyond the attempt to understand "the cultural significance of individual events" as they manifest themselves presently. They go on to ask the counterfactual question of what caused these events to be as they are when they could have been otherwise. As we will see shortly, it is only when a testable hypothesis is offered in response to this counterfactual question that the event in question (and by extension empirical reality itself) becomes "rational." In sum the soft sciences look at empirical reality in terms of *Sinn*, uniqueness/individuality, and variation/change. Having looked at the defining characteristics of empirical reality from the perspective of the soft sciences, the discussion turns to describing the type of conceptual tools that are capable of capturing this dimension of reality.

Weber notes, in very strong terms, that the conceptual tools of the hard sciences (in all their variety) are incapable of capturing empirical reality as understood by the soft sciences: "The sciences of concrete reality . . . attempt to establish a kind of knowledge which is necessarily unattainable if we accept the perspective of the nomological sciences: knowledge of *concrete reality*, knowledge of its invariably qualitative properties, those properties responsible for its peculiarities and its uniqueness" (RK, 57). The reason why this type of knowledge is "necessarily unattainable if we accept the perspective of the nomological sciences" is made clear when we look at the logical instruments (or type of concept), used by the hard sciences. Speaking of the nomological sciences, Weber notes:

> The definitive logical *instrument* of these disciplines is the use of concepts of an increasingly universal *extension*. For just this reason, these concepts become increasingly empty in content. The definitive logical products of these disciplines are abstract *relations* of *general validity* (laws). Their *domain* is that set of problems in which the *essential features* of phenomena—the properties of phenomena which are worth knowing—are identical with their *generic features*. (RK, 56f.)

In other words, the "logical instruments" (or concepts) employed by the hard sciences look at many, many different individual phenomena and seek to identify those features that are common (or generic) among the variety of individuals. The goal is to come up with those "generic features" that can be applied to as many of the observed individuals as possible. The more individuals that the "generic features" applies to (or the more "extensive" the concept), the more scientific the finding. This description of the nomological sciences is then contrasted with the sciences of concrete reality:

> The definitive logical instrument of these disciplines is the formation of relational concepts which are increasingly rich in *content*. In consequence, they are of increasingly limited *extension*. Their definitive products—if they can be said to have any conceptual character at all—are concrete *substantive concepts* of universal—or, as we may say, "historical"—*significance*. Their domain is found wherever the essential features of phenomena—that is those which we regard as worth knowing—are not exhaustively described by a classification of phenomena under some generic concept: that is to say, wherever concrete reality *as such* is the object of our interest. (RK, 58)

In contrast to the hard sciences, which seek to identify uniformity in the face of apparent diversity, the soft sciences seek to identify diversity in the face of apparent uniformity. Among a multitude of phenomena that can be labeled "capitalism," Weber argues that there is something unique about the capitalism that emerged in seventeenth-century northwestern Europe. For Weber this type of capitalism is of scientific interest not because of the generic features that it shares with the capitalism "that appeared in China, India, Babylon, the ancient [Greco-Roman] world and the Middle Ages" (PESC, 16), but because of the features that are unique to it and set it apart from other seemingly similar phenomena. Consequently, the concepts employed by the soft sciences are of very limited extension because they apply to a very particular phenomenon and attempt to describe it in as much detail as possible so that its unique features can be highlighted. In light of the two passages cited above we can draw the following comparison between the soft and hard sciences:

Table 4.1.

	Domain of Knowledge	Type of Concept	Logical Product
Hard Sciences	Generic features of group phenomena.	Concepts devoid of content but aspiring to universal extension because they apply to groups of phenomena.	Concepts of universal validity.
Soft Sciences	Unique features and cultural significance of concrete phenomena.	Concepts rich in descriptive content but of limited extension because they apply to only one individual in reality.	Concepts of historical significance.

From this comparison and contrast of the nomological sciences and *Wirklishkeitswissenschaften*, it is almost self-evident why knowledge of meaning, unique individuals, and variation "is necessarily unattainable if we accept the perspective of the nomological sciences." The nomological sciences (of which the hard sciences are one type) do not have the conceptual tools to study empirical reality as this term is understood by the soft sciences. But this obviously does not mean conceptual tools cannot be constructed to attain scientific knowledge of meaning, uniqueness, and variation. Weber notes:

> Consider a science which *treats* "personalities" or "actions" and their "motives" as *intrinsically* impervious to analysis because, from the point of view of *this* science, such an analysis would serve no valuable theoretical purpose. This circumstance alone is certainly not sufficient to eliminate such a discipline from the domain of the "objectifying" sciences. The biological concept of the "cell" in its relation to the concepts of physics and chemistry, has exactly the same properties. (RK, 137)

The point that Weber is making is self-evident—and needs to be detailed because of its self-evidence. All of the conceptual tools, experimental techniques, and equipment that are known to physics fail to provide even a high school textbook description of the phenomenon known as a "cell" in biology. It was not until biology came up with conceptual tools, experimental techniques, and observational equipment that were unknown to physics (and will forever remain foreign to physics) that we had a scientific understanding of "cell." Not only is this the case when we compare physics to biology or biology to chemistry. The exact same is the case within physics. All the conceptual tools, experimental techniques, and equipment known to classical physics fail to provide even a high school textbook description of a phenomenon known as "quantum mechanics." A set of concepts, experiments, and technology that is unknown (and will forever remain foreign to classical physics) had to be invented before we could come to have a scientific description of quantum phenomena.

The powerful hold of the "will-to-believe of naturalistic monism" can be vaguely glimpsed by considering the fact that physics has yet to come up with a theory that can explain physics (i.e., the relationship between classical, relativity, and quantum). By extension, the physical-natural sciences have yet to come up with a theory that explains the physical-natural sciences (i.e., the relationship between physics, chemistry, biology, and perhaps psychology). Yet we find positivists both in the hard and soft sciences advocating the position that the soft sciences should learn to mimic the conceptual tools and methods of the hard sciences if they are to become proper sciences. If we turn to history as a guide, instead of vague, semiconscious sentiments, we learn that a science becomes a science when it invents a novel set of conceptual, methodological, and experimental tools. This novelty in its turn allows us to see phenomena in empirical reality that were always present but remained hidden because of the underdeveloped and primitive condition of the existing sciences. In short just because the hard sciences do not have the conceptual tools to conceptualize the qualities, sentiments, and values on the one hand, the *Sinn*, individuals, and variation on the other hand does not mean that such tools cannot be constructed. And just as importantly it does not mean that the reality to which these concepts refer does not exist. On the contrary, careful analysis of the type of concept that is unique to the soft sciences reveals that the "concept" is far more than what the hard sciences assume. In other words the soft sciences make a unique contribution to developing a more scientific understanding of "concept." Similarly the scientific conception of "logical products" and the scientific domain of knowledge are radically altered when the unique contributions of the soft sciences are taken into account.

In addition to the nomological sciences and the sciences of concrete reality, Weber is aware of another group of sciences that do not belong to either of these two types—the mathematical sciences. Weber notes that the mathematical sciences employ concepts that are very different from those employed by either of the other two types of science. The latter two types of science need empirical reality in order to construct their concepts; the mathematical sciences can construct their concepts without any regard for empirical reality. Speaking of the concept of "pseudospherical space," Weber notes that "the empirical nonvalidity of pseudospherical space is irrelevant to the 'correctness' of such a construction" (RK, 190). Just because mathematical concepts do not meet certain criteria that are shared by the hard and soft sciences does not mean that the type of concept that is used by mathematics is an invalid type of concept. The validity of the type of concept used by mathematics depends on the type of knowledge that mathematics aims at. The only thing that needs to be demonstrated is that the type of concept that mathematics employs is able to reach the goal that mathematics has set for itself. Similarly, just because the soft sciences employ a type of concept that has certain unique characteristics distinguishing it from the concepts of the mathematical and hard sciences does not put the soft sciences outside the pale of science. In sum, Weber's reflections show that there are three types of science: (1) mathematical sciences, (2) nomological sciences, and (3) sciences of concrete reality. Generally speaking, the soft sciences belong to the sciences of concrete reality and the hard sciences belong to the nomological sciences.

In addition to describing the difference between the different types of science, Weber notes that there are different types of concepts that can be employed in scientific inquiry. Most of his attention is focused on drawing a distinction between two types of concepts: (1) "simple class or generic concepts (*Gattungsbegriffe*)"and (2) "the quasi-generic (*Gattungsmassigen*) *ideal-type*" (OSS, 100). In light of his discussion on the different types of science, we can safely assume that Weber is aware of a third type of concept, that is, the "mathematical concept." Speaking in the most general terms, the conceptual tool of the hard sciences is the *Gattungsbegriffe* (generic concept) and the conceptual tool of the soft sciences is the ideal-type which is a quasi-generic concept, *Gattungsmassigen.* The two different kinds of concepts are distinguished by the particular task that each of them seeks to accomplish. The generic concept (also called the "class concept") aims to identify "a complex of traits which are common to many phenomena" and "disregard the *meaning* of the component parts of the concept" (OSS, 100). Consequently the generic concept is employed by those sciences that are primarily concerned with identifying "a complex of traits" that are shared by many different individuals so that the individuals can be classified according to the types of traits that they share: "The more it is a matter of the simple classification of events which appear in reality as mass phenomena, the more it is a matter of class concepts" (OSS, 101).

In contrast: "The greater the event to which we conceptualize complicated historical patterns with respect to those components in which their specific *cultural significance* is contained the greater the extent to which the concept—or system of concepts—will be ideal-typical in character" (OSS, 101). The cultural significance of phenomena is captured not by explicitly making clear "the class or average character but rather the unique individual character of cultural phenomena" (OSS, 101). The ideal-type is a type of concept that is specifically designed to capture the cultural significance of unique individuals. The relationship between uniqueness, significance, and the ideal-type is captured in Weber's definition of the ideal-type as "a mental construct for the scrutiny and systematic characterization of individual concrete patterns which are significant in their uniqueness" (OSS, 100).

From this discussion it is apparent that the concepts that are employed by mathematics do not come under the heading of either "generic concepts" or the "ideal-type." Mathematical concepts are a type of their own because the end that mathematics aims for is different from the end of both the hard and soft sciences. Mathematics is not concerned with the generic features of empirical reality (as are the hard sciences). Nor is it concerned with the unique features of empirical reality (as are the soft sciences). It cares as little for "historical significance" as it does for "universal validity." The ultimate goal of mathematics is to provide a scientific description of pure possibility in all of its possible variety without any concern whether or not that possibility did, does, or will ever attain in empirical reality. Since the end of mathematical knowledge is different from the end of the hard and soft sciences, mathematics will have to employ concepts that are adequate to its chosen end. For Weber the soft sciences and hard sciences have as much authority to label mathematical concepts as being "invalid" as the mathematical or hard sciences have to label the ideal-type "invalid"—they have no authority whatsoever. The validity of a particular type of concept is determined by the end that a particular of science pursues—and the particular science has complete autonomy to determine its own end independent of the pressure, prejudices, and peculiarities of all the other sciences. That much having been said, a scientific conception of "concept" cannot be constructed from the vantage point of any one science—just as the scientific conception of "matter" cannot be constructed from the vantage point of just classical, relativity, or quantum physics. The scientific conception of "matter" can only be put forward after the three branches of physics have had their input and the differing perspectives have been brought into relationship with each other. Similarly, the scientific conception of "concept" can only be put forth after the mathematical, hard, and soft sciences have had their say. We have seen that Weber's work contains an implicit typology of science along with an implicit typology of concept. We have also seen that he explicitly links a particular type of concept with a particular type of science, based on the particular end (or domain of knowledge) of the particular science. We can summarize these linkages as follows:

Table 4.2.

Type of Science	Domain of Knowledge	Type of Concept
Mathematical Sciences	All the possible features possibly imaginable in any possible reality.	Mathematical concept
Hard Sciences	Generic features of observed phenomena in empirical reality.	Class or Generic concept
Soft Sciences	Unique features and cultural significance of concrete phenomena in empirical reality.	Ideal-type

With this table in mind, we now turn to Peirce's semiotics to help us better understand Weber's conception of the relationship between the type of science, its domain of knowledge, and the type of concept that it employs.

PEIRCE'S SEMIOTICS AND WEBER'S TYPOLOGY OF CONCEPTS

In chapter 3 we saw that Peirce's phenomenology brings clarity and precision to Weber's typology of experience. Now we will see that Peirce's semiotics brings clarity to Weber's typology of the "domain of knowledge" and the typology of concepts. The fact that Peirce's semiotics plays a clarifying role in Weber's discussion of the types of objects that can be scientifically investigated and the types of concepts that are used to study the objects is not only unsurprising but also something to be expected from both a Weberian and Peircean perspective. From the Weberian perspective, we saw that there is an intimate relationship between Weber's critique of the positivist and intuitionist conception of experience and their respective positions on the place of concepts in scientific inquiry. For Weber the faulty positions of these two groups on the role of concepts in scientific inquiry are directly related to their faulty (conflated) understanding of experience. Consequently, there is an intimate relationship between Weber's types of experience and types of concepts. In Peirce's case the relationship between phenomenology and semiotics is even more explicitly clear. For Peirce, while phenomenology is an inquiry into "experience" as experienced by psychical consciousness, semiotics is an inquiry into "experience" as experienced by intellectual reason. The central concern of phenomenology is to answer the question: "What is experience?" The central concern of semiotics, as understood by Peirce, is to answer the question: "How is thought about experience possible?" Peirce's summary answer to this question is: "We think only in signs. These mental signs are of a mixed nature; the symbol part of them are called concepts"

(2.302). Consequently, wherever there is thought, there are signs. These signs are composed of different elements—and the one element that is most familiar is "concept" (or the "symbol part" of the sign).

The most basic unit of semiotics is the triadic "sign"—consisting of the representamen (or sign-vehicle), object, and interpretant. Peirce notes that the phenomenological categories of Firstness, Secondness, and Thirdness apply to the characteristics and activities of intellectual reason as much as psychical consciousness. In other words, the categories apply as much to semiotics as phenomenology. Consequently, to answer the question "How is thought possible?" requires looking at the relationship between signs and the categories of Firstness, Secondness, and Thirdness.

The following description of a sign gives us the starting point of a stick-figure picture of Peirce's conception of the sign: "Anything which determines something else (its *interpretant*) to refer to an object to which it itself refers (its *object*) in the same way, the interpretant becoming in turn a sign and so on *ad infinitum*" (2.303). Summarily stated, a sign is made up of three elements: representamen, object, and interpretant. A note of caution—sometime Peirce uses the word "sign" to refer to the triadic relationship as a whole, and sometimes he uses it to refer to one of the three elements in the sign (the representamen). To avoid confusion, the term "sign-vehicle" will be used when referring to one of the three parts of a sign. Putting Peirce's verbal description into diagrammatic form, we come up with the following picture of a sign (figure 4.1).

We can take the verbal and diagram description of the sign and restate it in terms of Peirce's categories—using Peirce's own words: "A *Sign*, or *Representamen*, is a First which stands in such a genuine triadic relation to a

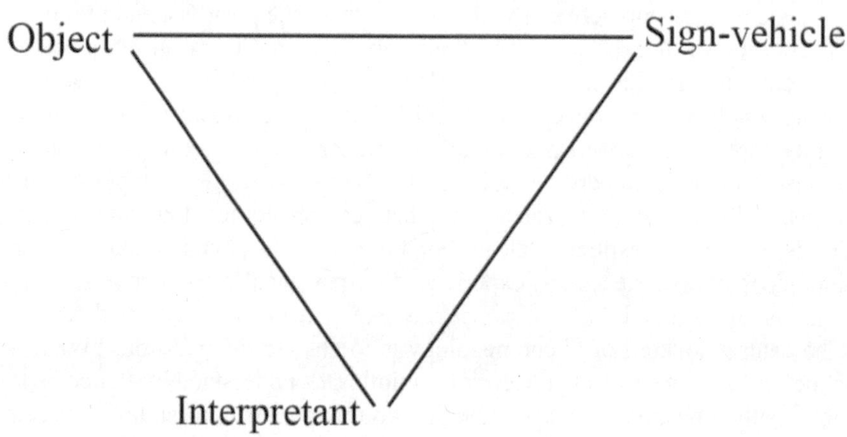

Figure 4.1.

Second, called its *Object*, as to be capable of determining a Third, called its *Interpretant*, to assume the same triadic relation to its Object in which it stands itself to the same Object" (2.274). The categories as applied to semiotics can be summarized as follows:

- First: Sign-vehicle (representamen)
- Second: Object
- Third: Interpretant

Here it must be kept in mind that we are dealing with First, Second, and Third at the level of Thirdness (or philosophy) because we are talking about the characteristics, functions, and dynamics of the sign in intellectual reason. Studying the sign as brute fact would mean looking at First, Second, and Third at the level of Secondness (ideoscopy or the special sciences). Studying the sign as possibility would mean dealing with First, Second, and Third at the level of Firstness (mathematics). Peirce's description of each of the three parts of the sign (i.e., sign-vehicle, object, and interpretant) overlaps with Weber's insights on the "domain of knowledge" of the difference sciences and the typology of concept. In order to appreciate this overlap we have to first look at the technical vocabulary introduced here by looking at Peirce's "trichotomy of signs" more closely.

Peirce notes:

> Signs are divisible by three trichotomies; first, according as the sign in itself is a mere quality, is an actual existent, or is a general law; secondly, according as the relation of the sign to its object consists in the sign's having some character in itself, or in some existential relation to that object, or in its relation to an interpretant; thirdly, according as its Interpretant represents it as a sign of possibility, or as a sign of fact or a sign of reason. (2.243).

The first of the three trichotomies looks at the sign "in itself" without any concern for the relationship of the sign to an object or an interpretant. At this level "a Sign may be termed a *Qualisign*, a *Sinsign*, or a *Legisign*" (2.244). As a Qualisign, a sign "is a quality which is a Sign. It cannot actually act as a sign until it is embodied; but the embodiment has nothing to do with its character as a sign" (2.244). The terms "red" and "ten" are examples of Qualisigns. Both of these terms refer to a quality—the first to the quality of "redness" and the second to the quality of "tenness." In both cases the terms by themselves "cannot actually act as a sign." To become an actual sign they have to be embodied in (something like) "red apple" or "ten apples." But the character of a Qualisign as a sign is not in the least affected by its embodiment or lack thereof. Stated in simpler terms, Qualisigns are Firstness as a sign of pure, unbounded possibility, quality, freedom, and spontaneity.

In contrast to a Qualisign, which is pure quality or possibility as a sign, a Sinsign "is an actual existent thing or event which is a sign" (2.245). Here we are looking at actual brute facts in the universe as signs, the facts of "trees," "atoms," "gravity," "politics," "capitalism," "religion," "stock market," "pilgrimage," etc. Each of these is a fact in the universe and hence a Sinsign. As an "actual existent thing or event," the Sinsign shares some characteristics with the Qualisign and at the same time it has unique, distinguishing characteristics. Since an "actual existent thing or event" is only "so thorough its qualities," all Sinsigns, "involve a qualisign, or rather, several qualisigns" (2.245). But at the same time, the Sinsign involves a "peculiar kind" of Qualisign—only those Qualisigns that are capable of "being actually embodied" (2.245). In sum, while Qualisigns are the types of signs that give us the ability to think about Firstness, Sinsigns are signs that make it possible for us to think about Secondness.

In addition to pure quality and brute facts being signs, habit-law[3] can also be a sign—Peirce calls this latter type of sign a Legisign. Peirce notes that a Legisign "is not a single object, but a general type which, it has been agreed, shall be significant" (2.246). Being a "general type," the Legisign is not a "fact" or "existent" but something that shapes "facts" and "existents" which are its "Replica": "Every legisign signifies through an instance of its application, which may be termed a *Replica* of it" (2.246). Peirce goes on to illustrate the relationship between a Legisign and a Replica through an example: "Thus, the word 'the' will usually occur from fifteen to twenty-five times on a page. It is in all these occurrences one and the same word, the same legisign. Each single instance of it is a Replica. The Replica is a Sinsign. Thus, every Legisign requires Sinsigns" (2.246). Similar to the fact that a peculiar type of Qualisign is embodied in a Sinsign, a peculiar type of Sinsign (i.e., the Replica) is contained within a Legisign, and it is through the Replica-Sinsign that the Legisign is instantiated. Peirce emphasizes that the characteristic of "significance" is something that is peculiar to the Legisign. A Qualisign or Sinsign becomes significant only by virtue of some habit-law. So while a Legisign needs a peculiar type of Sinsign to be instantiated, the Replica-Sinsign also needs a Legisign: "Nor would the Replica be significant if it were not for the [habit]-law which renders it so" (2.246). While Qualisigns make it possible for us to think about Firstness, and Sinsigns make it possible for us to think about Secondness, Legisigns are the types of signs that make it possible for us to think about Thirdness, that is, habit-law, generality, regularity, etc.

In chapter 2 we noted that Huff sees remarkable resemblance between Weber's notion of imputation and Peirce's conception of abduction. It will be too much of a stretch to claim that there is a similar resemblance between Weber's typology of concepts and Peirce's trichotomy of sign-vehicle. Nonetheless, Peirce's trichotomy of sign-vehicle does confirm and clarify a number of Weber's key insights. First and foremost, we see that there are different types of concepts—and each type is useful for a particular purpose. If (for example) a claim is made that only the generic concept is a genuine concept or that only the Sinsign is a real sign-vehicle, then actual, real, and important dimensions of empirical reality are closed off from scientific inquiry. While the generic concept cannot account for individuals and uniqueness in the universe, the Sinsign cannot account for generality and relationship. Secondly, Weber's description of the ideal-type concept does not make any sense without an appreciation of the "nomological" and "mathematical" elements that are embedded in it. Similarly if one does not recognize the peculiar type of Sinsign that is part of the Legisign and the peculiar type of Qualisign that is part of the Sinsign, one has not understood or read Peirce properly. In short, Weber and Peirce teach us to simultaneously recognize uniqueness, difference, and relationship between the different types of concepts/sign-vehicles. Thirdly, Weber's discussion of the concept and Peirce's discussion of the sign-vehicle bring us face to face with almost the exact same conundrum—the gap between meaning and significance on the one hand and validity and existence on the other. Weber described the logical products of the hard sciences as concepts devoid of content but of universal validity, and the logical products of the soft sciences as concepts rich in content and of historical/cultural significance. Peirce, for his part, describes the Replica-Sinsign as an existent fact that is devoid of significance and the Legisign as a general habit-law which "it has been agreed, shall be significant" (2.246). But by showing how a Legisign cannot be instantiated without a Sinsign (the Replica) and how the attribute of "significance" cannot be conferred on a Sinsign in the absence of a Legisign, Peirce is pointing the way to bridging the gap between validity/existence and significance/meaning.

Peirce's second trichotomy of sign describes the relationship of the sign-vehicle to the Object. At this level "a Sign may be termed an *Icon*, an *Index*, or a *Symbol*" (2.247). Summarily stated, an icon "is a sign which refers to the Object that it denotes merely by virtue of characters of its own . . . whether any such Object actually exists or not" (2.247). An index "is a sign which refers to the Object that it denotes by virtue of being really affected by that Object" (2.248). And a symbol "is a sign which refers to the Object that it denotes by virtue of a [habit]-law . . . which operates to cause the Symbol to be interpreted as referring to that Object" (2.249). In other words, an icon is a sign that merely suggests, an index is a sign that forces, and a symbol is a sign that shapes.

Going back to Weber's text, we can say that Ben Franklin's print shop and Richard Baxter's sermons, as examples of the "'spirit' of capitalism" and the "Protestant ethic" respectively, can be studied as icons, indices, or symbols. If (for example) Baxter's sermons are treated as an icon, it will mean that they are looked at merely in terms of the possibilities, feelings, qualities, and sentiments that they suggest. An example of iconic treatment of objects is found in Weber's perceptive remarks on the contrast between the concluding passages from Dante's *The Divine Comedy* and the concluding passages from Milton's *Paradise Lost*. After citing passages from *Paradise Lost*, Weber notes: "Every person immediately feels that this powerful expression of the grave Puritan turning-to-the-world—that is, the valuing of life's matter-of-fact activities as a task—would never have been expressed by a medieval writer. Yet this turning-to-the-world is just as incompatible with Lutherism, as is apparent, from example, in the hymns of Luther and [the poet and composer] Paul Gerhardt" (PESC, 47). Similarly, an iconic treatment of Baxter's sermons will certainly produce a "vague sense" of the "significant differences" (PESC, 47) between the possibilities, feeling, qualities, and sentiments underpinning the "Protestant ethic" and those underpinning Catholic or even Lutheran piety. Such a treatment is solely concerned with describing the "vague sense" conveyed by the sermons, without any regard for the existence of the object (or the actual existence of the sermons). In the most general terms, mathematicians and artists treat objects as icons or paint an iconic picture of the universe. Iconic symbols are one kind of degenerate symbol that Peirce calls "*Abstract Symbol*, whose only Object is a character" (2.293). In more technical language, looking at objects in the universe as icons is an iconic use of symbols that makes the objects degenerate symbols to the first degree.

In addition to an iconic treatment of Baxter's sermons, they could be treated indexically. Peirce notes: "An Index is a sign which refers to the Object it denotes by virtue of being really affected by that Object" (2.248). While an Icon retains its character as a sign even if its Object does not exist, an Index "is a sign which would, at once, lose the character which makes it a sign if its object were removed, but would not lose that character if there were no interpretant" (2.304). The indexical treatment of Objects leads to a strictly "objective" perspective—describing those characteristics that the objects themselves possess. The so-called "universal laws of nature" are deduced from observation of Objects as if there is no observer or as if the observer has no particular perspective or value-ideas shaping his observation. The goal of this "objective" observation is to identify those elements in the Indexical sign that it shares with other Indexical signs that are affected by or denote the same Object. Peirce calls it a "*Singular Symbol* whose Object is an existent individual and which signifies only such characters as that individual may realize" (2.293). An indexical treatment of Baxter's sermons means that the investigator is interested in the sermons as factually existing and identifying "only such characters" that these sermons themselves embody. This is a "just the facts, please" treatment of the sermons. The dates on which they were given, the different places at which they were given, the different topics that they covered, the number of people who heard the sermons in person, which sermons were put in writing (and in which publications) to reach a wider audience, etc. A study of Baxter's sermons in these terms would be an indexical treatment of the sermons. Treating objects as indices is an indexical use of the symbol that makes the objects degenerate symbols to the second degree.

The characteristic that sets the symbol apart from the icon and index is the fact that in order for a symbol to be a sign, all three—sign-vehicle, object, and interpretant—must be present: "A *symbol* is a sign which refers to the Object it denotes by virtue of a [habit]-law, usually an association of general ideas, which operates to cause the Symbol to be interpreted as referring to that Object" (2.249). The Symbol "would lose the character which renders it a sign if there were no interpretant" (2.304). The Symbol is a type of sign that leads the interpretant to refer to the same Object that it itself denotes. Turning the interpretant's attention toward the Object that it denotes is a crucial function of the Symbol because it is only through the Object that the Symbol can be perceived. Peirce notes that since a symbol is a type of general "that which is general has its being in the instances which it will determine. There must, therefore, be existent instances of what the Symbol denotes" (2.249). Consequently, while the Symbol itself is neither a fact nor an existent, it shapes and determines facts and existents. Just as an Index has a peculiar type of Icon inside of itself, the Symbol contains a peculiar type of Index: "The Symbol will indirectly, through a sort of Index, be affected by those instances [that instantiate it]: and thus the Symbol will involve a sort of Index, although an Index of a peculiar kind" (2.249). It should also be noted that since the Symbol is "a general type or [habit]-law . . . [it] is a Legisign. As such it acts through a Replica" (2.249). A symbolic treatment of Baxter's sermons would mean that the general habit-law of which these sermons are an instance would have to be identified. Furthermore, the manner in which these sermons shaped and determined their interpretants would also have to be described. This is the treatment that Weber focuses on in his work. A crucial part of Weber's argument (perhaps the most crucial) is that the sermons were genuine symbols that affected the sentiments, actions, and thoughts of their audience, thereby becoming a cause for the emergence of the "'spirit' of capitalism." Given the fact that a Symbol contains an Icon and Index inside of itself, the symbolic treatment means that the symbolic aspects of the sermons will have to be described in addition to the iconic and indexical aspects, not in their stead. (And we see that Weber does not spare himself the trouble of touching upon the iconic and indexical aspects of the sermons both in the body of the text and in the endnotes.)

Having looked at Peirce's second trichotomy of signs—the relationship between sign-vehicle and object—we turn our attention to looking at the clarifying role that it plays for Weber. Earlier in the chapter we saw that a typology of concepts is implicit in Weber's work. He identifies the mathematical concept, generic concept, and ideal-type concept as three different types of concepts. From the perspective of Peirce's semiotics there is no difference between the three types of concepts if they have been understood and put to use by an investigator—at this level all three function as symbols, in contrast to icons or indices. But at another level there is a clear difference between the mathematical and generic concepts on the one hand and the ideal-type concept on the other. The former two are examples of degenerate symbols while the ideal-type is an example of a genuine symbol. The mode of relationship of the mathematical concept to the object and the generic concept to the interpretant renders them degenerate (rather than genuine) symbols. Since the mathematical concept has no relationship to either an object or an interpretant, in terms of Peirce's semiotics, it is an iconic use of the symbol. Since the generic concept aspires to provide a description of objects without any concern for the role of the interpretant, it is an indexical use of the symbol. Simply stated, since the ideal-type establishes a relationship between sign-vehicle, object, and interpretant, it can be called a symbolic use of the symbol. But this simple statement will have to be unpacked in light of Peirce's observation that "while the complete object of a symbol, that is to say, its meaning, is of the nature of a law, it must *denote* an individual, and must *signify* a character. A *genuine* symbol is a symbol that has a general meaning" (2.293). A "genuine" symbol is composed of a general habit-law and denotes a particular individual for whom that habit-law signifies a character. Even as a heuristic device, the ideal-type comes very close to being a genuine use of the symbol because it is an attempt to conceptualize what (for example) the "'spirit' of capitalism" or the "Protestant ethic" meant for a particular group of people, in a particular place, at a particular time. The foregoing discussion can be summarized in the following manner:

Table 4.3.

Weber's Typology of Concepts	Peirce's Typology of Symbol	In Terms of Peirce's Semiotics
Mathematical Concepts	Abstract symbol	Iconic use of the symbol and a degenerate symbol in the first degree
Generic Concepts	Singular symbol	Indexical use of the symbol and a degenerate symbol in the second degree
Ideal-type	Symbolic symbol	Symbolic use of the symbol and a genuine symbol

There is another characteristic of the symbol that distinguishes it from the icon and index—"Symbols grow. They come into being by development out of other signs, particularly from icons, or from mixed signs partaking of the nature of icons and symbols" (2.302). With respect to "growth" Peirce has something very specific in mind: "A symbol, once in being, spreads among the peoples. In use and in experience, its meaning grows. Such words as *force, law, wealth, marriage*, bear for us very different meanings from those they bore to our barbarous ancestors" (2.302). To put this into Weber's terms, with the passage of time there has been a "variation of meaning" in "capitalism"—this term means something very different to the civilized moderns living in postindustrial society than it did to their semibarbarous early modern ancestors and barbarous medieval and premedieval ancestors. Neither the iconic symbol nor the indexical symbol is capable of capturing this "growth." Only a genuine symbol or the ideal-type has the ability to conceptually capture both the process of "growth" and the causes of the "growth." It is only with the help of this knowledge that one can offer a testable hypothesis as to why the growth took place in the particular direction that it did when it could have gone in a different direction. At this level only a genuine symbol can conceptually capture change and variation—the icon and index are incapable of allowing for change and variation in empirical reality, let alone capturing.

In chapter 2, we saw Weber putting for the evidence to demonstrate that the soft sciences "employ the category of causality in its full meaning" (RK, 196), while the hard sciences employ it in its truncated and degenerate form. Similarly, in light of Peircean semiotics we can expand Weber's argument and state that the soft sciences "employ the category of concepts in its full meaning." The concepts employed by the hard sciences are, by (Peirce's) definition, "degenerate concepts." Since by (Weber's) definition, causal interpretation and conceptual mastery of empirical reality are two of the defining characteristics of scientific knowledge, using "degenerate concepts" in scientific inquiry can only produce a "degenerate" scientific picture of empirical reality. Peirce's semiotics allows us to clearly see that abstract symbols (or mathematical concepts) and singular symbols (or generic concepts) are "degenerate symbols" with respect to the empirical reality in which the cultural activity of scientific inquiry is being carried out. In the present context, suffice it to say that every symbolic symbol (or ideal-type) contains an abstract and singular symbol inside of it. (Hence Weber's observation that discovering "laws"—or describing empirical reality in terms of singular symbols—cannot be the end of the soft sciences. It is an intermediate, necessary means for attaining the end.) While every symbolic symbol presupposes the existence of abstract and singular symbols, the latter two can be constructed without reference to a symbolic symbol. At the same time that Peirce's semiotics gives Weber the tools to describe his typology of concepts with greater precision, it also allows us to see more clearly how Weber's reflections on concepts and concept formation contribute to a more scientific understanding of concept—just as his reflections on "the full meaning" of causality contribute to a more scientific understanding of causality.

FROM THE WEBERIAN TO THE PEIRCEAN INTEGRATION OF BELIEF, ACTION, AND THOUGHT

Peirce's semiotics not only allows us to see Weber's typology of concepts with greater clarity and precision, it also allows us to better appreciate the depth and sophistication of Weber's reflections on the relationship between belief, action, and thought/values. In chapter 2 we saw that Weber's analysis of human action and social structures showed that wherever we find patterned action that is "objectively" observable, "subjective" beliefs and value-judgments are inevitably present. This conclusion on Weber's part was the result of a detailed logical analysis of patterns of social action. Peirce undertakes a similarly detailed logical analysis of human thought and comes to the conclusion that scientific thought is not possible in the absence of the sentiment of belief. In other words just as Weber's inquiry into the constituent elements of human conduct showed it is composed of belief, action, and thought, Peirce identifies the exact same three elements of being constituent elements of human thought (especially the species of "scientific thought").

We saw earlier that for Peirce, thought takes place only through signs and that every sign is made up of three distinct elements, that is, the sign-vehicle, object, and interpretant. Peirce notes: "If a Sign[-vehicle] is other than its Object, there must exist, either in thought or expression, some explanation or argument or other context, showing how—upon what system or for what reason the Sign represents the Object or the set of Objects that it does" (2.230). In the most general terms the "system" or "reason" that establishes a relationship between a sign and an object is "habit." Looking at the issue another way, Peirce notes that thought is composed of three elements: (1) ideas, (2) connections between ideas, and (3) the habit of establishing connection between ideas. When we look at what Peirce means by "habits of thought" or the habit establishing connection between ideas, we find that he explicitly identifies it with "belief": "A belief is itself a habit of the mind by virtue of which one idea gives rise to another" (7.354). A close analysis of "thought" reveals that it is much less like a "thing" that rests somewhere in the mind and much more like a process that the mind is constantly engaged in. Insofar as this process displays some degree of consistency, it is a sign of belief. Referring to an "idea" or a "thought" Peirce notes:

> It does not consist in anything which is present to the mind, but in an habitual connection among the things which are successively present. That is to say, it consists in ideas succeeding one another according to a general rule; but not in the mere thinking of this general rule, nor in the mere succession of ideas one

> upon another, nor in both together. A thought must therefore be a sign of a belief; but is never the belief itself. (7.355)

Given the fact that thought is a process (a pattern of intellectual action) and belief is a necessary prerequisite for displaying a pattern of action, an individual must be in a state of belief if she is displaying the habit of thought. In (7.355) Peirce carefully distinguishes thought from belief—we cannot confuse thought with belief because thought is not belief but only a sign of belief. The fact that thought is a sign of belief can be demonstrated by looking at the different roles that thought and belief play in relationship to action:

> When a person is said to act upon a certain belief the meaning is that his actions have a certain consistency; that is to say, that they possess a certain intellectual unity. But this implies that they are interpreted in the light of thought. So that even if a belief is a direct motive to action it still is a belief only because that action is interpretable again. And thus the intellectual character of beliefs at least are dependent upon the capability of the endless translation of sign into sign. (7.357)

In broad terms, belief brings consistency to action and thought brings interpretability to action. Just as consistency and interpretability cannot be conflated, similarly belief and thought cannot be conflated. Even though they are two unique individuals, thought and belief are intimately related: "A thought which is not capable of affecting belief in any way has no signification or intellectual value at all. If it does affect belief it is then translated from one sign to another as the belief itself is interpreted" (7.357). In other words, just as belief brings consistency to action, it brings consistency to thought (which is a type of action). And just as thought brings interpretability to action, it also brings interpretability to belief (which is also a type of action). In these reflections Peirce has demonstrated that belief has an intellectual (or cognitive) element, and thought (as well as action) has a sentimental (or intuitive) element. This is similar to the earlier observation that all signs have a symbolic element attached to them even though the sign itself may be an icon or an index.

Everything that has been said in this section is part of Peirce's theory of signs "as to what *must* be the character of all signs used by a 'scientific' intelligence, that is to say, by an intelligence capable of learning by experience" (2.227). From the foregoing discussion, it is clear that belief is a crucial element in all scientific thought. In addition to belief, Peirce identifies the sentiments of hope and charity as the necessary but insufficient conditions for scientific thought (this point will be detailed in chapter 7). Consequently, when we look at scientific investigation as praxis we find that it is the result of a dynamic relationship between belief, patterned action/conduct, and thought. Peirce's analysis of the constituent elements of scientific investigation has reached the same conclusion as Weber's analysis of the constituent elements of social reality such as "Christianity" or "the state." Weber's sociological analysis of human action comes to the conclusion that empirical reality is composed of beliefs, patterns of actions, and thoughts. Peirce's philosophical analysis of "scientific intelligence" comes to the conclusion that belief, action, and thought are essential preconditions for the scientific study of empirical reality. Putting Weber and Peirce together, we can draw the conclusion that, in the absence of belief, it is neither possible to engage in social action of any kind nor conduct scientific inquiry of any kind into social action (or any other aspect of empirical reality).

PEIRCE'S SEMIOTICS AND THE TYPES OF IDEAL-TYPE

In the previous two sections we have seen how Peirce's semiotics brings clarity to Weber's typology of concepts and Weber's reflections on the relationship between belief, action, and thought. In addition to these points, there is another point on which Peirce's semiotics plays a clarifying role for Weber. This latter point is with respect to the ideal-type not just as it is one type of concept among others but as a category that contains many different types inside itself. Peirce's ideas can bring clarity to Weber's reflections on the ideal-type on two different but related issues.[4] First, Peirce's semiotics gives Weber the conceptual framework to coherently categorize the multitude of ideal-types that is found in his writings—in other words, we will be able to construct a typology of ideal-types from Weber's work, using Peirce's semiotics. Second, Peirce's writings give Weber the language to describe "the common epistemological foundations" of these different types of ideal-types. Peirce's ideas will not only bring coherence to Weber's ideas on the ideal-type but also help us to begin to see the logical relationship between Weber the methodologist and Weber the social scientist.

In addition to constructing ideal-types to conduct his inquiry, Weber also conducted an inquiry in order to gather data that would help him to construct ideal-types. Referring to Weber's *Economy and Society*, Kalberg notes: "This three-volume, 1500-page treatise must be viewed as Weber's attempt to systematize the knowledge acquired from his case studies into ideal-types. All are designed to aid sociologists engaged in comparative-historical work. However, owing to the poor organization and unfinished character of E&S, this aim is not always apparent" (Kalberg, 1994, 90). Beyond "poor organization" and "unfinished character" of the book, the lack of a systematic framework to categorize the plethora of ideal-types contained in *Economy and Society* is a major factor for obscuring one of its main objectives. In the absence of this framework, *Economy and Society* appears to contain disjointed investigations of topics ranging from religion, law, politics and economics—all carried out with the aid of conceptual tools that seem to be arbitrarily thrown together. Consequently, as Hekman notes, in the secondary literature on the ideal-type, one often comes across the assessment that "Weber's concept is inconsistent, fraught with logical contradictions and incapable of dealing with the various levels of social scientific analysis" (Hekman, 1983, 120).

One can get an idea of the sheer quantity of ideal-types in Weber's work by looking at the unit of analysis that Weber uses in his sociological investigations: the social action of individuals. The social action of individuals is made up of two elements: objectively observable behavior and subjectively intended motives. The well-known ideal-types for human behavior constructed by Weber are: traditional action, value-rational action, means-end rational action, and affectual action. The ideal-types of human motives are: practical rationality, value rationality, theoretical rationality, and formal rationality. Weber goes on to note that all social action takes place within certain social structures. The investigator has to not only identify this structure but also determine its type. Let's assume the inquiry has something to do with the way human beings behave in different legal systems. An ideal-type of a "legal system" would have to be established, to distinguish it from "economic," "political," and other types of structures in society. Once that is done, the inquirer has to identify the basis of the legitimacy of the structure. The legitimacy of a structure can be based on one of four types: affectual attachment, value-rational commitment, religious sanction, or legal-rational legitimacy. The social action of the individual is not merely the result of an individual acting within a particular structure, the actor-structure relationship exists within a wider context. Depending on a variety of factors, the behavior of an individual in a legal setting will be affected differently. Consequently, the investigator must also take this factor into consideration. In technical terms these factors are "sociological foci for action" which are the "social contexts that demarcate constraints upon and opportunities for action" (Kalberg, 1994, 39). *Economy and Society* is full of ideal-types of these sociological loci, divided into the two broad categories of status groups and universal organizations (i.e., the household and neighborhood). It is difficult enough to keep track of the different types of ideal-types that have been mentioned in this example. An attempt to identify any relationship among even this relatively small number of ideal-types appears to be a futile exercise. Keeping in mind that the ideal-types mentioned here are taken from about seventy-five to a hundred pages of *Economy and Society*, one can well imagine what is contained in the other fourteen hundred pages.

Weber's work is an excellent example of constructing ideal-types of human action, social structures, and the different factors that facilitate or restrict human action in particular structures. But it is a very poor example of a description of the different types of ideal-types, the structure that links them to each other, and their relationship to each other. The first contribution that Peirce's semiotics can make to Weber's discussion of ideal-type is to provide Weber with a framework that allows him to present his ideal-types in a coherent and systematic manner. Peirce has a system of categorization that contains a list of 144 different types of signs. If not in its entirety, at least a part of this list will be able to bring coherence and systematization to the laundry list of ideal-types in Weber's work.

The second point on which Peirce's semiotics can bring greater clarity to Weber's reflections on the ideal-type is in terms of articulation. Hekman notes that Weber's methodological reflections on the ideal-type provide "the common epistemological foundations" (Hekman, 1983, 125) for constructing a wide variety of ideal-types. Hekman focuses on four particular types: sociological ideal-types, historical ideal-types, ideal-types of action, and ideal-types of motives. While each of the four captures a different aspect of reality, they share a common epistemological foundation. When looking for the cause of this commonality we do not need to go further than the definition of "culture" provided by Weber. As noted above, Weber defined culture as "a finite segment of the meaningless infinity of the world process, a segment on which *human beings* confer meaning and significance" (OSS, 81). The importance of relationship and continuity is evident in the very definition of culture. Culture is the result of an unending process of constructing (and deconstructing) relationships. The ideal-type is the conceptual tool that is capable of capturing not only the relationships that are formed in order to bring culture into existence but also the process by which the relationships are formed (and deformed and reformed). Translated into the language of Peirce's semiotics, the ideal-type is a legisign that not only (conceptually) captures relationships, it is also a symbol that embodies relationships and is capable of describing how relationships develop in the universe. In even more technically precise terms, the symbol is the embodiment of Thirdness, a sign of Thirdness, and a description of the growth of Thirdness.

Since culture is composed of many different types of relationships, many different types of ideal-types are needed to conceptualize different types of relationships. Examples of two different types of relationships captured by two different types of ideal-types are: (1) between the subjective and objective aspects of a phenomenon and (2) between the subject conducting the inquiry and the object being studied. In the first case "the category of facts from which an ideal-type is constructed must be seen in its 'cultural significance,' that is, in the meaning which the social actors bestow on their actions" (Hekman, 1983, 122). At this level the ideal-type conceptualizes the *Sinn* that an actor bestows upon her action by establishing a relationship between the objectively observed behavior of the actor with her subjectively intended motives. In the second case: "The synthesis of selected aspects of . . . culturally significant facts must take cognizance of the relationships of meaning and significance logically compatible with the theoretic interest informing the investigation and the logic of the social actor's concepts" (Hekman, 1983, 122). For Weber it is the ideal-type that bridges the subjective-objective and subject-object dichotomies; for Peirce the task of bridging dichotomies is performed by the triadic sign (or symbol).

The centrality of the ideal-type in Weber's science is no less than the centrality of the symbol in Peirce's philosophy because the importance of relationships in empirical reality is the same for both thinkers. But on one important point there is a difference between the two: Weber's discussion on the ideal-type implies that, in spite of their bewildering variety, all ideal-types share a "common epistemological foundation"—but he does not detail this point. In contrast, Peirce's work carefully describes not only this common foundation but also how one gets from this foundation to each of the different types of symbols. This is the second point on which Peirce brings greater clarity to Weber's reflections on the ideal-type. Peirce gives Weber the precise philosophical language to describe the "common epistemological foundations" on which all ideal-types are constructed. This evidences that Peirce's semiotics can play the same role in clarifying Weber's reflections on concept formation that von Kries's probability theory played in clarifying Weber's conception of causality.

At this point in the discussion it would be worthwhile to bring together comments that have been made intermittently about the soft sciences using the categories of causality and concepts in their full and genuine meaning and the hard sciences using these two categories in their degenerate and truncated forms. The logical and methodological evidence has been presented during the course of the discussion. The implications of the full and genuine use of causality and concepts in scientific inquiry are the following. To begin with, the investigator can recognize phenomena in empirical reality that cannot be recognized when using the truncated and degenerate forms. Both Weber and Peirce demonstrate that sentiments, qualities, and value-judgments/commitments/ideas are part and parcel of empirical reality, as much as (if not more than) atoms, cells, and planets are part of empirical reality. Secondly, empirical reality can be studied in terms of significance/meaning, individual uniqueness, and developmental variation as much as (if not more than) in terms of objectivity, universality, and immutability. The first point reveals that empirical reality has aesthetic and ethical dimensions. The second point makes it possible to study certain characteristics of empirical reality that remain beyond the reach of the hard sciences. This is the outcome of moving from a truncated and degenerate understanding of the causality and concepts of the hard sciences to the full and genuine conception of the soft sciences. Peirce's phenomenology and semiotics help us to better understand Weber's insights on these two points. But in one sense we have only begun our study. As we move forward and look at how Peirce's pragmaticist logic clarifies Weber's reflections on "conceptual apparatus" we will be able to see that the soft sciences employ scientific logic "in its full meaning" while the hard sciences can only employ it in its truncated and degenerate form. The implications for our understanding of what "science" is and what it can (and cannot) do are far-reaching.

NOTES

1. A question might emerge regarding the approach used to link Weber and Peirce. The first part of the chapter will "reconstruct" Weber's position on concept formation—bringing together Weber's reflections on the issue from different places. The second part of the chapter will describe Peirce's phenomenology and semiotics. And the final part of the chapter will link the two. The observation by Pierre Bourdieu is an apt summary of the approach that is being used in this chapter (and chapter 5)—the only difference is that we will substitute the term "Peirce's pragmaticism" where Bourdieu uses the term "symbolic interactionism."

> One need only bring these various passages together and give them their full significance to derive, in a first break with Max Weber's explicit methodology, a representation of the relations between religious agents that may be termed *interactionist* (in the sense in which we speak today of *symbolic interactionism*). If this is a view of things that has to be read 'between the lines", this is because, so far as we can see, the intellectual tools Weber had at his disposal prevented him from forming a clear awareness of the principles he was applying (at least intermittently) in his research and consequently from setting them to work in a methodical and systematic fashion. The fact that it would not be difficult to extract the explicitly stated principles of a theory of symbolic interaction from Weber's theoretical writings makes the reformulation of Weberian analysis in the language of symbolic interactionism all the easier and, it would seem, all the more legitimate. (Bourdieu, 1987, 121)

2. By treating human culture as a symbol system and linking the meaning of the conduct of the actors to the way that the meaning is understood in that particular context, Weber is anticipating the trend of semiotics anthropology that emerged in the late 1970s. Milton Singer (1984) and Michael Silverstein at the University of Chicago combined the semiotics of Peirce and Roman Jakobson to offer tools for anthropological inquiry that linked the analysis of meaning that social actors invested in their activities with their cultural surroundings, rather than deriving the meaning of the action of the actors from some disembodied theory of language. This was a departure from the methods of inquiry based on the works of Levi-Strauss and Saussure which gave a place of privilege to the methods and theories of the investigator in determining the meaning of cultural activity. Mertz (2007) describes the emergence of semiotics anthropology and surveys the different trends within it since its emergence.

3. Peirce uses the term "law" in a number of different ways. Sometimes he uses it in the positivistic sense—as a brute fact in the universe. For the positivists, law as brute fact is the "cause" of observed events. In more than a few places Peirce offers "law" as an example of Secondness. At other places he describes "law" not as a brute fact but as a general tendency or habit. Here "law" is an example of Thirdness. I will use the term "habit-law" to avoid the confusion, where I think Peirce is using "law" in the sense of Thirdness.

4. John G. Gunnell (2007) notes that Wittgenstein offers reflections that are very close to Weber's description of the ideal-type:

> Wittgenstein might be construed as identifying something approximating Weber's image of ideal-types when he spoke of the manner in which philosophy, or any meta-practice, might approach its conventionally constituted subject matter. Wittgenstein suggested that what was required in giving an account of a *Lebensform* and *Weltbild*, that is, what might be called cultural objects, was a "perspicuous representation" or "sketches of a landscape" that "produces just that understanding which consists in 'seeing connections.'" . . . The kind of representation that he sought would be accomplished by "inventing *intermediate cases*" that determined "the way we look at things." The subject matter consisted of "language games" embedded in various social practices and forms of life, but he recommended and saw the necessity of creating, second order language games that would be "set up as *objects of comparison*" and were "meant to throw light of the facts . . . by way not of similarities, but also dissimilarities." . . . Such a model would, again, be "an object of comparison . . . a measuring rod; not as a preconceived idea to which reality must correspond." . . . It would be difficult to imagine a better account of the kind of thing Weber talked about as an ideal-type. (Gunnell, 2007, 84f.)

While Wittgenstein could give Weber the philosophical language in which to describe the ideal-type, the following discussion will argue that Peirce's class of signs gives Weber not only the language in which to describe the ideal-type but also a system to coherently categorize them.

Chapter Five

Conceptual Apparatus and the Logic of Scientific Inquiry

In addition to the concept of causality and the significance of concept formation, the third point on which Weber sees no difference between the hard and soft sciences is the conceptual apparatus used in scientific inquiry. Weber notes that while the different types of sciences employ different types of concepts, there is no difference in the "conceptual apparatus" that is used to construct the different types of concepts and conduct the investigation itself. Weber describes the three different types of apparatus as: (1) intuition and imagination, (2) abstraction and generalization, and (3) comparison and contrast. For Weber this conceptual apparatus not only makes it possible to construct concepts, it also describes the different stages of scientific inquiry. In this second function the term "conceptual apparatus" can be interpreted as "scientific method" or the "logic of scientific inquiry." In this chapter we will see how Peirce's logic brings clarity to Weber's understanding of the scientific method.

Before getting into the details of the relationship between Weber's understanding of the scientific method and Peirce's logic, a brief look at Peirce's description of the relationship between reasoning and logic would be helpful. He describes the goal of "reasoning" in these terms: "The object of reasoning is to find out, from the considerations of what we already know, something else which we do not know. Consequently, reasoning is good if it be such as to give true conclusions from true premises, and not otherwise. Thus, the question of the validity is purely one of fact and not of thinking" (5.365). While the goal of reasoning is to reach "true conclusions from true premises" it must depend on logic to distinguish "true premises" from "false premises" and the "valid" line of reasoning linking premises to conclusions from "invalid" line of reasoning. Peirce describes the relationship between reasoning and logic by noting that

> Reasoning involves deliberate approval of one's reasoning; and approval cannot be deliberate unless it is based upon the consideration of the thing approved with some idea of how such a thing ought to appear. Every reasoner, then, has some general idea of what good reasoning is. This constitutes a theory of logic: the scholastics called it the reasoner's *logica utens*. (2.186)

Logic is the specialized science in philosophy specifically concerned with a close and detailed study of the reasoning processes and seeks to establish objective criteria that separate valid reasoning from invalid reasoning. In the simplest terms the science of logic studies the different ways knowledge about the unknown can be inferred on the basis of the known. Before Peirce two types of inferences were known, deduction and induction—Peirce named a third abduction (or retroduction). This third type of inference plays a critical role in bringing clarity to Weber's reflections on the scientific method.

Peirce notes that there is a close relationship between his semiotics and his logic: "Logic, in its general sense, is . . . only another name for *semiotics*" (2.227). This suggests that Peirce's logic can play the same clarifying role vis-à-vis Weber's "conceptual apparatus" that his semiotics played in clarifying Weber's insights on concept formation. Peirce notes that scientific inquiry is composed of three different types of inferences—abduction, deduction, and induction. This chapter will begin by bringing the three elements of Weber's "conceptual apparatus" into relationship with Peirce's three modes of inference. This exposition will show that Weber and Peirce are offering practically the exact same description of the scientific method, using different terms. After getting a coherent account of the three elements that any scientific inquiry is composed of, the discussion will move on to looking at the third of Peirce's trichotomy of signs—the three different ways that the interpretant of a sign can represent the sign. This will set the stage to look at the relationship between understanding, explanation, and interpretation.

INTUITION AND THE GENERATION OF HYPOTHESES

Weber's pointed critique of the intuitionist position that concepts are not needed in scientific inquiry provides a good starting point for beginning the discussion of his conception of the scientific method. The intuitionists claim that they can dispense with concepts in scientific inquiry due to the (alleged) fact that the investigator can have unmediated knowledge of the phenomenon that he is studying because both the investigator and the phenomenon belong to the domain of *Geist*. Weber notes that the position on the intuitionists is based on a basic confusion: "They confuse the question of the psychological *conditions* responsible for the genesis of knowledge with the completely different question regarding the logical 'content' of knowledge and its empirical 'validity'" (RK, 170). In this dense and important passage, Weber identifies three distinct but intimately related steps in scientific inquiry: (a) the genesis of knowledge claims (i.e., determining the origins of knowledge claims), (b) the articulation of knowledge claims (i.e., determining the "logical 'content' of" the knowledge claims), and (c) the testing of knowledge claims (i.e., determining the "empirical 'validity'" of knowledge claims). The apparatus that generates, articulates, and tests hypotheses is the same in all scientific inquiry. As we will see, for Weber intuition and imagination are the source of all scientific hypotheses, abstraction and generalization are used to articulate them, and comparison and contrast are used to test their validity. The intuitionists confuse the generation of hypotheses with their testing and go on to claim that there is no need for comparison and contrast in socio-cultural inquiry. The positivists are so obsessed with issues relating to the articulation and testing of hypotheses that they completely lose sight of the fact that something very different is at work in their origin and generation.

In very explicit and strong terms, in a variety of different texts and contexts, and in both his methodological and sociological writings, we find Weber repeatedly drawing our attention to the fact that the origin of scientific hypothesis, everywhere and always involves the same "subjective" element—the intuition and imagination of the investigator. Weber describes the role of intuition in generating scientific hypotheses in these words: "For the psychological genesis of an hypothesis in the mind of an historian [an intuition] is certainly of eminent importance. In fact, it is quite indispensable. Not only no valuable history, but no knowledge of any sort, has ever been 'produced' by a mere manipulation of 'observations' and 'concepts'" (RK, 177f.). It is not just the case that no valuable knowledge in history or the soft sciences has been the product of the "manipulation of 'observations' and 'concepts'"; the same is true of the hard sciences. Weber notes that "the really great advances in knowledge in mathematics and the natural sciences" are products of "intuitive flashes of imagination" (LCS, 176). An individual lacking a vibrant intuitive imagination may very well become a great "clerk or . . . technical official" (SV, 136), but such a person will never be a great scientist. Great scientists have the ability to "divine" novel scientific hypotheses by means of an "'intuitive' gift" (LCS, 176) on the basis of known observations and familiar concepts. If we make the allowance of considering "inspiration" the same as "intuition," Weber notes:

> Inspiration plays no less a role in science than it does in the realm of art. It is a childish notion to think that a mathematician attains any scientifically valuable

> results by sitting at his desk with a ruler, calculating machines or other mechanical means. The mathematical imagination of a Weirstrass is naturally quite differently oriented to meaning and result than is the imagination of an artist and differs basically in quality. But the psychological processes do not differ. Both are frenzy (in the sense of Plato's "mania") and "inspiration." (SV, 136)

Peirce uses the term "abduction" or "retroduction" to describe the "intuitive flashes of imagination" and "inspiration" that Weber identifies as the origin of all scientific discoveries. He notes that as important and indispensible as deduction and induction are in scientific inquiry, neither the origin nor the conclusion of scientific inquiry can be traced back to them:

> Observe that neither Deduction nor Induction contributes the smallest positive item to the final conclusion of the inquiry. They render the indefinite definite; Deduction Explicates; Induction evaluates: that is all. . . . Yet every plank of its advance is first laid by Retroduction [Abduction] alone, that is to say, by the spontaneous conjectures of instinctive reason; and neither Deduction nor Induction contributes a single new concept to the structure. Nor is this less true or less important for those inquiries that self-interest prompts. (6.475)

In very strong terms Peirce identifies abduction as both the origin of all new knowledge claims and as the ultimate ground on which "the final conclusion of the inquiry" is based. While they play an indispensable role during the course of the inquiry, induction and deduction neither initiate scientific inquiry nor bring it to a conclusion. For Peirce abduction is the alpha and omega of scientific inquiry.

In the broadest terms abduction refers to the ability of "guessing right." When we seek to identify the source of this ability, we are led to human instincts. Peirce notes that human beings display the characteristic of guessing right when faced with a novel, surprising phenomenon: "Nature is a far vaster and less clearly arranged repertory of facts than a census report; and if men had not come to it with special aptitudes for guessing right, it may well be doubted whether in the ten or twenty thousand years that they may have existed their greatest mind would have attained the amount of knowledge which is actually possessed by the lowest idiot" (2.753). For Peirce the amount of knowledge attained by human beings about the natural world in the last ten to twenty thousand years is the most compelling evidence for the fact that they have a special aptitude for guessing right. If human beings did not have this ability it is highly unlikely that the brightest among them could know more than the dumbest among them has actually known. The reasoning behind the position is the following. When faced with a novel phenomenon, there is slightly less than an infinity of possible explanations. If a human being had to go through each of these possibilities one by one (either inductively or deductively), to eliminate the false explanations to arrive at the correct answer, it would take almost an eternity. The fact that the better minds among us arrive at the correct hypothesis from among the almost infinite theoretical possibilities, within a few guesses, suggests that we do have a "special aptitude for guessing right." Peirce goes on to note that it is "not man merely, but all animals" who display this characteristic of guessing right (2.753). Among other things, Peirce calls this characteristic a "divinatory power" (6.477) and a "natural instinct for guessing right" (7.220). For Peirce it makes no sense to posit that this characteristic is at work in the animal kingdom but plays no role in the human attempt to understand (and master) the natural environment. Ordinary common sense attributes the ability of chickens (and other species in the animal kingdom) to "guess right" to their instincts. In light of this commonsensical explanation, Peirce asks: "But if you are going to think every poor chicken is endowed with an innate tendency toward a positive truth, why should you think that to man alone this gift is denied?"(5.591).

For Peirce it is important to acknowledge the role of instincts and intuition in scientific inquiry. Based upon his study of the history of science, Peirce comes to the conclusion that all important scientific discoveries through the ages have been the result of instinctive conjectures. Peirce notes that "the well-prepared mind has wonderfully soon guessed each secret of nature is historical truth. All the theories of science have been so obtained" (6.476). In very strong terms he notes that "the existence of a natural instinct for truth is, after all, the sheet-anchor of science" (7.220). While instinctive, abductive guesses may be the "sheet-anchor of science," they do not make up the totality of science.

ABSTRACTION AND ARTICULATION OF HYPOTHESES

As important as intuition and imagination are in the generation of scientific hypotheses, they have their obvious limitations. Prior to taking a determinate form that can be communicated to others, the intuitive guess or hypothesis is a "value-judgment" that is "felt" by the investigator to be worthy of affirmation. Weber notes: "It is *determinateness of content* which removes the object of the value judgment from the sphere of that which is merely 'felt'" (RK, 183). The determinateness of content is lent to the intuition by articulating it in a conceptual language that is shared with others. Given the discussion on the different types of experience discussed in chapter 4, it must be emphasized that the conceptual form in which the feeling is expressed is not the same as the experience of the feeling itself; it is an abstraction from that experience. In other words, while an imaginative intuition is the indispensable condition for the origin of scientific knowledge, conceptual abstraction is the indispensable condition for the articulation and communication of this knowledge—this is as true for the soft sciences as it is for the hard sciences.

Weber notes that conceptual abstraction is required to construct the simplest of propositions whose empirical validity is asserted. Even simple statements such as "Peter is taking a walk" or "Mary is thinking about her parents" presupposes "logical operations as soon as it constitutes a 'proposition' the 'validity' of which—for *this* is the only relevant question—is to be established. Although these logical operations do not include the 'justification' of general concepts, they certainly do include their constant *use*" (RK, 169). Just as we employ abstractions and generalizations in everyday speech, "the 'validity' of which . . . is to be established," so is this the case with all scientific hypotheses. Weber notes that the abstractions and generalizations that are employed in the interpretation of human action and ideas by the soft sciences do not differ from those employed by the hard sciences:

> Consider the "empirical generalizations" which are used to verify "interpretations" of human action. They only appear to differ from the generalizations used to verify interpretations of concrete "natural processes." This appearance, which is unimaginably superficial, is a consequence of the following. Our imagination has been schooled in the world of our own everyday experience. In consequence, when we are engaged in "interpreting" human action, we often omit the explicit *formulation* of the content of our experience into "generalizations." This would be a "waste of time." Therefore we use these generalizations "implicitly." (RK, 170f.)

Weber acknowledges the fact that in the interpretation of human action there is a unique element of "self-evidence," but (unlike the intuitionists) this recognition does not lead him to the conclusion that we can dispense with abstraction and generalization—quite the contrary, it leads him to deepen the inquiry into the character of "self-evidence." Weber notes:

> Suppose we "understand" human action as determined by clearly conscious and intended "ends" and a clear knowledge of the "means" required for these "ends." It is incontestable that the degree of "self-evidence" attained by this sort of understanding is unique. Suppose we consider what this degree of "self-evidence" is based upon. It is obviously the following circumstance: the relation between "means" and "ends" is intrinsically accessible to a rational *causal account which produces generalizations*, generalizations that have the property of "nomological regularity." (RK, 186)

As we saw in chapter 2, "nomological regularity" in the soft sciences is expressed in "empirical rules." These rules are the product of abstraction and generalization. Speaking of the actor being observed by the investigator, Weber notes: "There is no rational action independent of the causal rationalization of certain aspects of reality which are conceived as objects and instruments that can be manipulated; that is, independent of the systematization of these elements into a complex of empirical *generalizations* which indicate what results are to be *expected* from a specific action" (RK, 186f.). For Weber it is obvious that "without the belief in the reliability of empirical generalizations, there could be no action based upon an estimation of the means required for an intended result" (RK, 187)—this is as true in the first-person case as it is in the second- and third-person cases. The only difference between the investigator and the actor is that the former constructs her conceptual scheme with more attention to detail and precision than the latter, but in the end it is a more precise exercise of what is commonsense experience. The investigator constructs "conceptual schemes" or "ideal-typical constructions" (RK, 189) that deduce the consequences of certain human actions "by presupposing strictly rational action" (RK, 189). In short, the articulation of a hypothesis cannot take place without the aid of abstract concepts and conceptual schemes. While there is "fundamental methodological distinction" between knowledge of abstract laws and generalizations on the one hand and knowledge of socio-cultural phenomena on the other, "the construction of a system of abstract and therefore purely formal propositions analogous to those of the natural sciences, is the only means of analyzing and intellectually mastering the complexity of social life" (OSS, 87).

When we look at Peirce's description of the role of deduction in scientific inquiry we see that it closely corresponds with Weber's description of the role of abstraction and generalization. Peirce notes that "a hypothesis adopted by abduction could only be adopted on probation and must be tested" (5.202). But before the hypothesis is tested, the hypothesis has to be clearly and coherently articulated in such a way that it makes certain predictions whose validity can be put to the test. This marks the second stage of scientific inquiry:

> This testing, to be logically valid, must honestly start, not as Retroduction [Abduction] starts, with scrutiny of the phenomena, but with examination of the hypothesis and a muster of all sorts of conditional experiential consequences which would follow from its truth. This constitutes the Second Stage of Inquiry. For its characteristic form of reasoning our language has, for two centuries, been happily provided with the name Deduction. (6.470)

Deduction prepares the way for the testing of the hypothesis in two different but related ways. The first step in deduction is "by logical analysis to Explicate the hypothesis, that is, to render it as perfectly distinct as possible" (6.471). In this step some of the implicit assumptions in the hypothesis are made explicit. The second step is to state the hypothesis in the form of a theorem with predictive power. In this second step the goal is to "trace out [the hypothesis's] necessary and probable experiential consequences" (7.203). While instincts and intuition are predominant features underpinning abduction, abstraction and generalization are the predominant features characterizing deduction. Peirce notes:

> Reasoning is of three types, Deduction, Induction, and Abduction. In deduction, or necessary reasoning, we set out from a hypothetical state of things which we define in certain abstracted respects. Among the characters to which we pay no attention in this mode of argument is whether or not the hypothesis of our premises conforms more or less to the state of things in the outer world. We consider this hypothetical state of things and are led to conclude that, however it may be with the universe in other respects, wherever and whenever the hypothesis may be realized, something else not explicitly supposed in that hypothesis will be true invariably. (5.161)

As noted above, deduction cannot be considered the "testing" stage of the hypothesis, it is the preparatory stage for the testing. This is due to the fact that "deduction . . . relates exclusively to an ideal state of things" (7.205)—"ideal" here meaning hypothetical, theoretical. But the ideal description produced by deduction is an essential step in determining the validity of the hypothesis in the real world. This is because the second stage of inquiry provides us with certain expectations with which the results of an inductive experiment will be compared so that an informed judgment can be passed on the soundness of the abductive hypothesis.

COMPARISON AND THE TESTING OF HYPOTHESES

The process of scientific inquiry does not end with the generation of a hypothesis in the investigator's intuitive imagination and its articulation in precise terms with the aid of abstract concepts. The process must move on to the stage of hypothesis verification. For Weber the verification of a hypothesis invariably requires comparison and contrast. While there are some distinguishing characteristics of hypotheses put forth by the soft sciences, this does not mean that no comparison and contrast are needed during the stage of hypothesis verification. For Weber causal interpretation, the only kind of interpretation with which scientific inquiry is concerned, invariably requires the use of comparison:

> It is not the exception but rather the rule, that interpretation proceeds by the use of *analogies*. That is, by introducing the "experiences" of *others* which are intentionally *selected* for the purpose of comparison. Therefore, an unconditional presupposition of interpretation is a certain amount of abstraction and analysis. This is not only a *necessary condition* for "interpretation." "Interpretation" must be analyzed and verified in this way in order to have the properties of clarity and precision which Gottl ascribes to it *a priori*. (RK, 162)

Weber posits that the verification of hypotheses in the soft sciences is also based upon the comparison of observed events with theoretically expected results—results derived from the conceptual manipulation of facts in a conceptual, interpretive scheme. He notes that in the soft sciences: "The given facts are compared with a possible interpretation—*an interpretive scheme. To this extent*, therefore, the role of such conceptual schemes [in the soft sciences] is related to the role which teleological interpretation plays in biology" (RK, 189).

The fact that comparison and contrast are indispensible for hypothesis verification in the soft sciences is made obvious by the very character of the ideal-type:

> Whatever the content of the ideal-type, be it an ethical, a legal, an aesthetic, or a religious norm, or a technical, an economic, or a cultural maxim or any other type of valuation in the most rational form possible, it has only one function in an empirical investigation. Its function is the comparison with empirical reality in order to establish its divergences or similarities, to describe them with the *most unambiguously intelligible concepts* and to understand and explain them causally. (MEN, 43)

The use of comparison in the verification of hypotheses allows us to not only judge the validity of the hypothesis, but also the soundness of the intuitive imagination of the investigator who generated the hypothesis in the first place. Speaking of the ideal-type, Weber notes:

> It is a conceptual construct (*Gedankenbild*) which is neither historical reality nor even the "true" reality. It is even less fitted to serve as a schema under which a real situation or action is to be subsumed in one *instance*. It has the significance of a purely ideal *limiting* concept with which the real situation or action is *compared* and surveyed for the explication of certain of its significant components. Such concepts are constructs in terms of which we formulate relationships by the application of the category of objective possibility. By means of this category, the adequacy of our imagination, oriented and disciplined by reality, is *judged*. (OSS, 93)

In the final analysis, the verification of the hypothesis takes place by comparing the hypothesis to events in empirical reality. Whatever other differences there may be between the socio-cultural and natural sciences, it is simply not the case that the latter must employ comparison and contrast to verify hypotheses, while the former can dispense with them.

A look at Peirce's description of the role of induction in scientific inquiry shows that it closely corresponds with Weber's description of the role of comparison and contrast in the testing of hypotheses. Taking his description of the process of scientific inquiry further, Peirce describes the third stage of scientific inquiry in these words:

> The purpose of Deduction, that of collecting consequents of the hypothesis, having been sufficiently carried out, the inquiry enters upon its Third Stage, that of ascertaining how far those consequents accord with Experience and of judging accordingly whether the hypothesis is sensibly correct, or requires some inessential modification, or must be entirely rejected. Its characteristic way of reasoning is Induction. (6.472)

While deduction reaches its conclusion with reference to an ideal state, induction undertakes experimental tests to see if the conclusion obtains in empirical reality. Described a little bit differently, induction is an inference that tests predictions that are to be expected from the deductive conclusions that have been abstracted from an abductive hypothesis. For Peirce this type of inference "is alone properly entitled to be called *induction*" (7.206). The entire inductive stage is characterized by comparison. It draws samples from the actual world, compares them with the expected/theoretical results (derived from the deductive stage) and checks to see if there is a correlation between the two. While induction helps us to reach probable conclusions about actual experience, Peirce emphasizes that inductive conclusions lead to only a very particular type of "positive knowledge." Peirce notes that induction offers "mere processes for testing hypotheses already in hand. The Induction adds nothing [to the hypothesis]. At the most it corrects the value of a ratio or slightly modifies the hypothesis in a way which had already been contemplated as possible" (7.217).

The similarities between Weber and Peirce are quite striking in terms of the logic of scientific inquiry.[1] The only difference between the two is that Peirce is using technically precise vocabulary to articulate his position. Weber's position is being articulated in the midst of a heated debate with the advocates of positivism and intuitionism which often took on the character of polemics. Consequently, much of what Weber has to say is what conceptual apparatus is not, rather than clearly stating what it is according to him. If we look beyond the confrontational and polemical elements in his writing, we see that Weber has a very coherent position. Translated into Peircean terms, the following correlation emerges between Weber's "conceptual apparatus" and Peirce's logic of scientific inquiry:

Table 5.1.

Weber's "Conceptual Apparatus"	Peirce's Logic
Intuition, imagination, and the generation of hypotheses.	Abductive inference
Abstraction, generalization, and the formulation of hypotheses.	Deductive inference
Comparison, contrast, and the testing of hypotheses.	Inductive inference

This is one way to render Weber's "conceptual apparatus" into Peirce's logic:

1. The observer's attention is drawn to something.
2. A series of phenomena is observed: p, p`, p" . . .
3. Against the claims of J. S. Mill, Peirce argues that we cannot suppose any "uniformity in nature." By "uniformity," we mean "regularity," and by that we mean that which we recognize to display some anticipated regularities or patterns of phenomena. It would be a petitio principii to suppose both that the pattern is "given out there" *and* that the pattern is what I will recognize *as* a pattern. Mill had it backwards. There is no passive induction. In order to "see" a pattern out there, I must first imagine what it might be and, then, probe the data (the series of phenomena) in light of what I imagine.
4. So, what I first imagine is the idea of a really possible regularity or pattern. For Peirce the process of imagination may be likened to an abductive inference. The abduction recommends that if the surprising phenomena illustrated a given regularity, x, then we would recognize it and account for what we have seen or experienced. The conclusion of the abduction is what we call a hypothesis.
5. Once offered, the results of abduction may then be tested through an inductive inference. For Peirce induction does not generate our idea of regularity; instead, it receives that idea as a hypothesis and examines a series of phenomena to test whether or not they confirm what we would have anticipated.
6. Peirce suggests that the three forms of inference, abduction, induction, and deduction could be compared and contrasted by identifying each one with a different ordering of Aristotle's classical syllogism:

    a. Abduction:

    - The surprising object S displays the predicates P1, P2, P3
    - A well-recognized kind of object, M, tends to have for its predicates P1, P2, P3 . . .
    - Hence S may be of the kind M.

    b. Induction:

    - The surprising object S may be of the kind M.
    - If it is M, S would tend to display the predicates P1, P2, P3 . . .
    - S displays the same predicates P1, P2, P3

- (The hypothesis appears confirmed with respect to this sampling.)

c. Deduction:

- A well-recognized kind of object, M, tends to have for its predicates P1, P2, P3 . . .
- (If) the surprising object S is of the kind M
- (Then) S will display the same predicates P1, P2, P3.

7. Thus:

   a. The scientific inquiry requires what some would call the "subjectivity" of an a priori hypothesis." It must be tested by what some would call the "objectivity" of induction.
   b. The process effectively attributes something analogous to such a subjectivity *to the world* as that "course or habit of action" that *would*, if we are right, display the phenomena P1, P2, P3 . . .

8. Restated in Weberian-like terms:

   a. A social phenomenon is observed.
   b. It appears to display some statistical uniformity, S (unknown)=(P1, P2, P3 . . .).
   c. To begin to identify (comprehend, grasp) the uniformity, we engage in a process of abductive reasoning that concludes when we hypothesize that S is of the kind M, which would mean that each member of the series displays actions expressing the corresponding motives of the actors (MP1, MP2, MP3) . . .
   d. To test the hypothesis inductively, we interview the actors and retrieve evidence of what tends to be a series of motives (M1, M2, M3). . . . We conclude that the hypothesis is most likely confirmed, since we can correlate most actions to what we would anticipate would be their corresponding motives (MP1, MP2, MP3) . . .

9. S therefore appears to be a course of action whose causal interpretation is most likely comprehended.[2]

In the previous three chapters we saw that Peirce's phenomenology brings greater clarity to Weber's typology of "experience" and that Peirce's semiotics brings greater clarity to Weber's typology of concept. Now we see that Peirce's logic brings greater clarity to Weber's typology of conceptual apparatus. This clarity in its turn allows us to build on a theme that has been emerging in the previous discussion. We have seen that the hard sciences employ the categories of causality and concept formation in a truncated and degenerate form, while the soft sciences employ these categories in their "full meaning." Along the same lines, the triadic logic of scientific inquiry that is explicit in Peirce's work and implicit in Weber's work shows us that the soft sciences employ the logic of science in its full and genuine meaning, while the hard sciences can only employ it in its truncated, degenerate form. The reason for this is simple and straightforward. Not only do the soft sciences recognize the role of subjective elements in scientific inquiry, they alone are sufficiently equipped to undertake a scientific investigation into the validity and significance of these elements. This statement can be formulated in even greater detail and precision. We have seen that for Weber wherever there is patterned action, social institutions, and human culture, always and everywhere aesthetic and ethical elements are involved. In the same way, we saw that for Peirce wherever there is human thought (especially scientific thought) sentimental beliefs are always present. Since scientific inquiry is a product of human thought, carried out in observable patterns of actions, institutionalized in universities, research labs, and the like in human society, and part and parcel of human culture, wherever there is scientific inquiry certain beliefs and value commitments are inevitably present. The soft sciences are not only able to recognize the role of these "subjective" elements that are a constituent part of the "objective" phenomenon called "scientific inquiry," they are especially qualified to study the unique characteristics, significance, development variation, and empirical validity of these elements—in the same way that among all the sciences physics is especially qualified to study atoms, quasars, and planetary systems. In short, the soft sciences employ the logic of scientific inquiry in its full, genuine meaning because they not only employ it during the course of scientific inquiry but also have the capacity to bring the logic of scientific inquiry under scientific scrutiny.

By this time in the discussion, it is clear that the soft sciences employ the categories of causality, concepts, and logic in the full and genuine meaning, while the hard sciences employ these categories only in their truncated, degenerate form. If this is the case, then it follows that only the soft sciences engage in scientific investigation in the "full meaning" of the term, while the hard sciences can only engage in scientific inquiry in a truncated, degenerate sense of the term. A look at Weber's reflections on the relationship between understanding, explanation, and interpretation in light of Peirce's third trichotomy of signs suggests that this is indeed the case.

THE VARIETIES OF MEANING AND INTERPRETATION

In chapter 4 we saw that Peirce's triadic semiotics and Weber's three-tier typology of concepts make it possible to conceptualize dimensions and characteristics of phenomena in empirical reality that remain beyond the reach of dyadic semiotics or a two-tier typology of concepts. Peirce's symbolic symbol and Weber's ideal-type make it possible to conceptualize the aesthetic and ethical dimension of reality and study empirical reality in terms of *Sinn*, uniqueness, and variation. These are dimensions and characteristics of empirical reality that always and everywhere remain inaccessible to iconic symbols and indexical symbols (i.e., mathematical concepts and generic concepts). On this point Weber and Peirce render a service to science that is akin to the invention of the microscope and telescope—vastly expanding the scope of reality that is accessible to scientific observation. It is obviously not the case that microscopic bacteria were not a part of empirical reality before Zacharias Janssen invented the microscope. Nor is it the case that certain planets came into existence only after Galileo invented the telescope. It is obviously the case that scientists did not have the adequate tools to observe these phenomena. In a similar vein, we can say that before Weber conceived the ideal-type and Peirce came up with the notion of the symbolic symbol, it is not the case that aesthetic and ethical elements were not a not a part of empirical reality, and empirical reality was devoid of unique individuals, dynamic variation, and significance. But it is the case that after the contributions of Weber and Peirce, conceptual tools are available that make it possible to study these dimensions and characteristics of empirical reality scientifically.

Given the intimate relationship between Peirce's semiotics and logic on the one hand and Weber's typology of concepts and conceptual apparatus on the other, we can offer a very confident abductive hypothesis that Peirce's triadic logic and Weber's three-fold conceptual apparatus have made a significant contribution to the understanding and practice of scientific inquiry. This contribution can be summed up in the following words: Just as their insights vastly expand the domain of empirical reality that is accessible to scientific investigation, Peirce's logic and Weber's conceptual apparatus vastly expand the type of scientific knowledge about empirical reality that can be attained. Summarily stated, the integration of the insights of Weber and Peirce shows us that scientific understanding, explanation, and interpretation can be attained of events in empirical reality. Just as important as it is in showing that these three different types of scientific knowledge can be attained about the same event, their insights also show the reflexive relationship among the three. We will begin with Weber's insights on this topic and then move on to Peirce.

For Weber the dichotomy between understanding and explanation is bridged by causal interpretation—the ultimate goal of scientific inquiry. When we look at Weber's detailed description of what is entailed in causal interpretation, it becomes clear why Weber insists that "Sociology must reject the assumption that 'understanding' and causal 'explanation' have *no* relationship to one another" (CIS, 157)—and the one place where "understanding" and "explanation" come into relationship with each other is in a "causal interpretation." Weber gives the following description of causal interpretation:

> A correct causal interpretation of a concrete course of action is arrived at when the overt action and the motives have both been correctly apprehended and at the same time their relation has become meaningfully comprehensible. A correct causal interpretation of typical action means that the process which is claimed to be typical is shown to be both adequately grasped on the level of meaning and at the same time the interpretation is to some degree causally adequate. (ES, 12)

This passage shows that a correct causal interpretation of "a concrete course of action" is composed of three different elements. The most basic element is that the investigator correctly apprehends the "overt action and the [subjective] motives" of the actor. (Here it is worth repeating that the "investigator" can refer to even first-person reflections on one's own feelings, actions, and thought.) This is the element of "understanding"—where the investigator "apprehends" or "understands" the meaning (*Sinn*) that the act has for the actor. We have already seen that Weber categorically rules out the possibility of attaining direct or unmediated understanding of even one's own acts, to say nothing of second- and third-person acts. In order to understand "a concrete course of action," it must be shown that the observed, concrete event in question is a particular instance of a "typical" course of action. This "typical" course of action is expressed in the form of an "empirical generalization" or ideal-type. This ideal-type helps the investigator to "explain" the action on two levels. First, it is able to identify the meaning of the act for the typical actor (rather than the actual, concrete actor). As we saw in chapter 3, meaning is the emergent property that results from bringing facts into relationship with values, or actions into relationship with motives. The second level of explanation is to show that the motives were "causally adequate"—or with all other things being equal, the motives increased the probability of the occurrence of the event. When the concrete event in question is shown to be a particular instance of a typical trend, the important intermediate step toward causal interpretation (i.e., "explanation") is fulfilled. Weber explicitly identifies the different types of ideal-types that are used to construct the explanatory scheme:

> Objectively correct rationality serves sociology as an ideal type in relation to empirical action; instrumental rationality as an ideal type in relation to what is psychologically understandable; the meaningful as an ideal type in relation to "meaningless" action. Through comparison with the ideal type, the causally relevant irrationalities (different on each level) can be established for the purpose of causal attribution. (CIS, 156f.)

Merely showing that the concrete event is a particular instance of a general pattern cannot be confused with causal interpretation because the most important step in the inquiry remains to be taken—causal attribution or causal interpretation. This means identifying the "causally relevant irrationalities" that account for the reason that the particular event conforms to (or deviates from) the general pattern. In order for the relationship of motives and actions in the concrete event to become "meaningfully comprehensible," the investigator has to provide a testable hypothesis identifying the reason for the conformity or deviance of the concrete event from the ideal-type. This would be the investigator's "interpretation" of the event. In the absence of this final interpretive step, the event falls well short of a "causal interpretation." In sum we can see that a causal interpretation is composed of three elements (or steps):

- Understanding or "correctly apprehending" the observed actions and subjective motives at work in the event. (Or answering the question: "What is the meaning of this concrete event for the actor?")
- Explanation or identifying the probabilistic typical, general trend under which this particular event can be categorized. (Or answering the question: "Which empirical generalization does this particular event instantiate?")
- Interpretation or the investigator adequately grasping the meaning of the event by offering a reason for the event's conformity to or deviance from the typical, general trend. (Or answering the question: "Why is this event as it is, when it could have been otherwise?")

From the foregoing discussion it is clear that each of the three elements (i.e., understanding, explanation, and interpretation) that collectively make up the process of causal interpretation make a uniquely valuable contribution to the process. In this particular description of causal interpretation, Weber emphasizes the importance of attaining "adequacy" both with respect to meaning and causality. With respect to meaning, Weber notes: "If adequacy in respect to meaning is lacking, then no matter how high the degree of uniformity and how precisely its probability can be numerically determined, [the event under investigation] is still an incomprehensible statistical probability, whether we deal with overt or subjective processes" (ES, 12). Even if in 100 percent of the cases an investigator observes that social actors in

Western culture tip their hats when passing by each other in the street, the act of "tipping hats" remains "an incomprehensible statistical probability" if the investigator fails to apprehend the *Sinn* that the act of tipping hats has for the actors who tip their hats. Conversely, even if the *Sinn* of an event is perfectly apprehended, the event becomes sociologically significant only when (and if) the event can be shown to be an instance of general trend: "On the other hand, even the most perfect adequacy on the level of meaning has causal significance from a sociological point of view only insofar as there is some kind of proof for the existence of a probability that action in fact normally takes the course which has been held to be meaningful" (ES, 12). Adequacy at the level of causality is attained when a testable hypothesis is offered about the degree to which and why this event conforms to or deviates from a known "typical" pattern of action. It is only when the particular event is analyzed as an instance of a general pattern that a hypothesis can be offered about the "subjective" motives that have caused the event to be as it is when it could have been otherwise—always keeping in mind that the "empirical generalizations" that capture the "typical" pattern of action can never be accorded causal agency.

Here Weber is showing us that as important as empirical generalizations, statistical probabilities, mathematical equations, etc., are in facilitating the explanation of an event, neither the *Sinn*, nor the cause, nor the reason of the event can ever be derived from these abstract mathematical entities (either collectively, individually, or in any combination). In the absence of adequacy at the level of meaning (*Sinn*), the particular event remains completely outside the domain of rationality even if there is 100 percent statistical correlation or mathematical validity. Conversely, as important as the understanding of *Sinn* is for the scientific understanding of an event, the validity of the causal factors (or motives) can no more be derived from mathematical laws than they can from the meaningfulness of the event. In the absence of a hypothesis about causal adequacy, the motives remain completely outside of the domain of validity even if the *Sinn* of the event has been apprehended completely (or affirmed with complete conviction). In sum, scientific inquiry (in the soft sciences) has absolutely no interest in two types of events. Firstly, events that cannot be brought into the domain of rationality—these are events whose *Sinn* is beyond scientific investigation or understanding. Secondly, events that cannot be brought into the domain of validity—these are events whose cause(s) remain beyond logical and empirical scrutiny.

As we noted in the introduction, Weber was acutely aware of the fact that his methodology of scientific inquiry was underpinned by "a *theory of 'interpretation,'* a theory which at this point is barely visible and has hardly been explored at all" (RK, 151). From the foregoing discussion we can identify some of the defining characteristics of the "theory of interpretation" (or hermeneutics) that Weber was looking for. It must be based on a triadic logic that is able to bring understanding, explanation, and interpretation into relationship. We have seen that Peirce's first trichotomy of signs clarifies Weber's typology of concepts; the second trichotomy clarifies Weber's reflections on the relationship of the concept to the object; and Peirce's triadic logic clarifies Weber's reflection on "conceptual apparatus." This suggests that Peirce's work contains the resources to help Weber in terms of hermeneutics also. At this point we turn to Peirce's third trichotomy of signs.

The third trichotomy of signs categorizes the sign according to the relationship of the sign-vehicle to its object as it is represented by the interpretant. The interpretant can represent the object as "a sign of possibility or as a sign of fact or a sign of reason" (2.243). If the interpretant is a Rheme, the sign will be a class-8 sign (a sign of possibility); if it is a Dicisign, it will be a class-9 sign (a sign of fact); and if it is an Argument it will be a class-10 sign (a sign of reason). Looking at Peirce's description of these three classes of signs, it becomes easy to see what Weber is driving at when he is drawing distinctions (and relationships) between understanding, explanation, and interpretation. Understanding is treating an observed phenomenon as a class-8 sign, explanation is treating it as a class-9 sign, and interpretation is treating it as a class-10 sign. We will look at each of these three classes of sign in turn.

In a very real sense there are some experiences that cannot be "talked" about, "described," or "thought" about—they can only be "seen" or "felt." The one phenomenon that fits this description is "feeling," which is "an instance of that sort of element of consciousness which is all that it is positively, in itself, regardless of anything else" (1.306). Class-8 signs are symbolic legisigns that represent feelings, which cannot be otherwise "talked" about, or "described." The reason that feeling cannot be talked about is because: "A feeling is immediate consciousness, that is, whatever of consciousness there may be that is immediately present, yet there is no consciousness in it because it is instantaneous. [This is because a] feeling is nothing but a quality and a quality is not conscious: it is a mere possibility" (1.310). In having an aesthetic experience, we are having an experience of a class-8 sign—a "feeling, the consciousness of which can be included in an instant of time, passive consciousness of quality, without recognition or analysis" (1.377). In making possible the experience of "mere possibility," class-8 signs are not just means to explore fantastical possibilities, they also make it possible to be receptive to real possibilities. Sheriff notes: "Art does not just give us the unreal, the pure play of form in our imagination; it gives us symbols that make us aware of more of our experience in the world of signs, more of what is for us real" (Sheriff, 1989, 90). Approaching experience by "feeling it" and "seeing it" rather than by "thinking about it" or "describing it" is the hallmark of art in all its various forms (i.e., poetry, painting, music, architecture, etc.).

For Weber an investigator has to have the skills of an artist when she is looking at socio-cultural phenomena because far more is involved in any socio-cultural phenomenon (let's say Franklin's print shop as an example of the "'spirit' of capitalism") than the objectively observable elements and material. It is not possible to give a scientific account of the print shop without the ability to "feel" and "see" the subjective belief and values that are among its constituent components along with its organizational structure, historical location, profit-and-loss statements, and other material elements. This means that the investigator has to understand the meaning, significance, and value of the print shop (as much as is possible, given the resources available to her today) in the same way that they were understood by Franklin. This part of the investigation reveals that a particular Calvinist conception of *Beruf* (vocation, inner calling) is part and parcel of the print shop. A description of the print shop in terms of the sentiment and values that it embodied would be a Rhematic interpretation of the sign (the print shop) in Peirce's terms or Weber's conception of *Verstehen* (i.e., understanding).

While art is an example of class-8 signs, propositional statements are examples of class-9 signs. Propositional statements are claims about facts in the world (in contrast to claims about possibilities in the world), and the meaning of these statements is determined by the validity of the claim. In representing a symbolic legisign as a class-9 sign, the interpretant is claiming that this sign actually exists in the world. Peirce notes that a Dicisign "is a sign which is understood to represent its object in respect to actual existence" (2.252). Describing class-9 signs, Sheriff notes that they are "propositions that are true or false; that is, they are signs of fact that pretend to be really affected by their object . . . and, as such, are subject to the tests of evidence and probability with which propositions are validated" (Sheriff, 1989, 106). A description of Franklin's print shop as a class-9 sign would require that the investigator present sufficient evidence to demonstrate that this particular print shop was a capitalist enterprise. The proposition that has to be tested in this context is "Franklin's print shop is an example of a capitalist enterprise in colonial America." The description of a phenomenon as a class-9 sign is "just the facts, please" type of description where those facts about the phenomenon are presented that are relevant to the investigative concern. The treatment of print shop as a Dicisign in Peirce's terms is explanation in Weber's terms.

In representing a symbolic legisign as an Argument (or a class-10 sign), the interpretant is claiming that the symbolic legisign is a sign of reason or habit-law in the universe. Peirce is using the term "law" in a very particular sense here—the sense is captured by showing the relationship of "law" to facts (class-9 signs) and possibility (class-8 signs). Peirce notes that "a Rheme is a sign which is understood to represent its object in its characters merely; that a Dicisign is a sign which is understood to represent its object in respect to actual existence, and that an Argument is a Sign which is understood to represent its Object in its character as Sign" (2.252). In other words, the Argument represents its Object as a Sign—as something that stands for something for someone. Given the fact that the same Sign could stand for many different things for different people and even different things for the same person at a different time, the Argument will also have to show why the Sign stands for this thing at this time for this person, when it could be otherwise. Here we can see the difference between Peirce's notion of "law" and the typical positivist/Newtonian conception of law. (It is for this reason that I have used the phrase "[habit]-law" in the previous chapters to distinguish Peirce's conception from the typical conception.) For Peirce a class-10 sign is an example of a "law":

> As *general*, the law, or general fact, concerns the potential world of quality, while as *fact*, it concerns the actual world of actuality. Just as action requires a peculiar kind of subject matter, which is foreign to mere quality, so law requires a peculiar kind of subject, the thought, or . . . the *mind*, as a peculiar kind of subject foreign to mere individual action. Law, then, is something as remote from both quality and action as these are remote from one another. (1.420)

A law is a "general fact" that is a sign of meaning as reason and it is the fact of its generality that sets it apart from quality and fact. The fact that "law" as it is used here could mean something different for someone else, or even for the same person under different circumstances, evidences that "law" as Thirdness or Argument means something radically different for Peirce than the typical definition of this term. A description of the print terms in terms of a class-10 sign means that the investigator has to offer a testable hypothesis about its "reason" or the "argument" that the print shop is making. This reason and argument cannot be separated from the beliefs and values that are a constituent part of the print shop.

There were many other print shops and capitalist enterprises in Franklin's own day (as well as before him and after him) that were virtually indistinguishable from Franklin's when viewed as class-9 signs (or in terms of explanation). Those phenomena that cannot be distinguished at the level of a class-9 sign become distinct when studied as class-10 signs. Given Weber's definition of capitalism (when considered as a class-9 sign), Franklin's print shop is as good an example of "capitalist enterprise" as the business activity of the Franciscans. And both of these are virtually indistinguishable from the activity of Blackbeard. In terms of "explanation," all three are examples of "capitalist enterprise" because all of them sought to minimize losses and increase profits. But the situation changes radically when the same phenomena are studied as class-10 signs. As class-10 signs all three are signs of "reason" in the universe and all three signs are making an "argument"—but they are pointing toward three very different conceptions of "reason" and making three very different "arguments." In the final stage of inquiry, the investigator has to put forth a testable hypothesis of the reason that the phenomenon embodies and the argument that it is making. This is done in the form on an interpretation. Stated in simple terms, the "reason" that the phenomenon embodies is arrived at by exploring the question: "Why is X as it is when it could have been otherwise?" The reason that Franklin's actions embodied (which he himself probably understood and could explain only vaguely) can be clearly and coherently expressed in the aftermath of a scientific investigation that delves into the question of why he chose to pursue profit in the way that he did when he could have chosen otherwise. A comparison of the reason behind Franklin's activity, and those of the Franciscans and Blackbeard, will reveal that his reason was very different from those of the other two. The goal of scientific inquiry is to distinguish the three types of reasons and the different arguments that they are making. In other words, the goal of interpretation in scientific inquiry is not to reach the Reason of the universe, but to discover the different types of reasons and arguments that are possible in the universe.

The discussion of Weber's conception of understanding, explanation, and interpretation on the one hand, and Peirce's conception of the interpretant as Rheme, Dicent Sign, and Argument, reveals striking affinity in the insights of the two thinkers. Given the close proximity of the two in terms of phenomenology, semiotics, and logic, the affinity at the level of hermeneutics is not all that surprising. Combining the language that they have used, we can say that the ultimate goal of scientific inquiry is for an investigator to put forth an Argument (in the form of a testable hypothesis) that answers the question "Why is X as it is, when it could have been otherwise?" In the absence of a testable hypothesis as an answer to this question, the phenomenon remains outside the domain of rationality. Events in the universe, as well as the universe itself, become rational only when and insofar as a causal interpretation is offered to account for the fact that the observed facts are as they are when they could have been otherwise. The following table summarizes the link between Weber and Peirce in terms of the types of knowledge that a scientific investigation can generate:

Table 5.2.

Weber's *Wissenchaftslehre*	**Peirce's Third Trichotomy of Signs**
Understanding	Studying phenomena as class-8 signs
Explanation	Studying phenomena as class-9 signs
Interpretation	Studying phenomena as class-10 signs

At this point we begin to see that while the soft sciences engage in scientific inquiry in the full sense of the term, the hard sciences can engage in scientific inquiry in only its truncated, degenerate form. The hard sciences cannot go beyond treating empirical phenomena as class-9 signs (as facts in the universe). In Weberian terms they cannot go beyond explanation. Not only can the soft sciences go beyond explanation, for Weber they must go beyond explanation and offer a causal interpretation of observed phenomena. We saw in chapter 2 that the hard sciences cannot offer a genuine causal account of events in the universe. When an apple falls to the ground, the logically correct statement is "The falling of the apple does not contradict our knowledge of the laws of gravity/science/nature"—this cannot be confused with a genuine causal account. Similarly, the hard sciences do not have the conceptual capacity to capture unique individuals, variation, and significance in/of empirical reality. The most they can do is indentify the generic features and invariable rules/patterns under which different groups of unique individuals can be grouped. A description of empirical reality solely in generic and invariable terms is a truncated, degenerate description of reality. Finally, the hard sciences cannot even theoretically imagine the question: "Why is X as it is when it could have been otherwise?" In other words it is not possible to offer an interpretation of observed phenomena, given the final ends that the hard sciences aim for.

At the end of chapter 4 we noted that a movement from the truncated, degenerate conception of causality and concepts of the hard sciences toward their full, genuine conception in the soft sciences expands the horizons of scientific inquiry. The full, genuine conception of causality and concepts makes it possible to recognize and conceptually master the aesthetic and ethical dimensions of empirical reality, and also makes it possible to study empirical reality in terms of individual uniqueness, dynamic variation, and cultural significance. At this point in the discussion we can say that in addition to expanding the horizon of scientific inquiry, the soft sciences contribute to the deepening of the inquiry by making it possible to ask certain investigative questions that cannot possibly be asked by the hard sciences. Neither the hard sciences nor the positivist or intuitionist conception of the soft sciences can tackle the question: "Why is X as it is, when it could be otherwise?" There is no phenomenon in the universe about which this question cannot be asked. And for Weber any phenomenon about which this question cannot be asked and an answer provided is utterly irrational.

At this point in the discussion we can identify the implications of employing the categories of causality, concepts, and logic in their truncated, degenerate forms in contrast to employing them in their full, genuine forms. Employing them in their truncated, degenerate form necessarily leads to either a dualist or reductionist conception of the hard and soft sciences. The intuitionists have a truncated, degenerate understanding of these three categories in the sense that they think that there is a segment of empirical reality where these categories simply do not apply. They reject the use of causal analysis, conceptual tools, and inductive logic in the study of socio-cultural phenomena. This position sets up an ontological and epistemological divide between the hard and soft sciences. The positivists employ the three categories in a truncated, degenerate form in a different sense. They reduce causality to causal equivalence, conceptual tools to generic concepts, and scientific logic to induction and deduction. This positions leads to studying socio-cultural phenomena with the same tools and methods used to study physical-natural phenomena—in other words reducing the soft sciences to the terms of the hard sciences. The unbridgeable divide between the hard and soft sciences produced by intuitionist dualism should not be confused with the reduction of the soft sciences to the terms of the hard sciences produced by positivist reductionism. But whatever difference there may be between intuitionist dualism and positivist reductionism at other levels, there is no difference between the two at the level of practical consequences. Two of the most important consequences are: (1) a pseudo-scientific conception of science and (2) an inevitable and irreconcilable conflict between the hard and soft sciences.

The truncated, degenerate conception of causality, concepts, and logic produces a pseudo-scientific conception of science for the same reasons that defining physics solely from the perspective of Newtonian physics produces a pseudo-scientific conception of physics. As has been noted previously, there is much more to physics (and much more to reality) than Newtonian physics is capable of even theoretically imagining. Any answer to the question "What is physics?" that does not take account of relativity and quantum physics would produce a pseudo-scientific conception of physics. Similarly any answer to the question "What are the hard-sciences?" that is constructed from the perspective of Newtonian-relativity-quantum physics alone, and does not take into account the uniqueness of biology and chemistry, would produce a pseudo-scientific conception of the hard sciences. Taking the line of reasoning further, we can say that the unique contributions that the soft sciences make to our understanding of the categories of causality, concepts, and logic have to be taken into account and brought into relationship with the contribution of the hard sciences in order to produce a scientific answer to the question "What is science?"

Moving from a pseudo-scientific conception of science to a scientific conception has far-reaching consequences for our understanding of the relationship between the hard and soft sciences. It is only by moving toward a scientific conception of science (in which input from all the different sciences is taken into account in answering the question "What is science?") that we can see the deep and intimate relationship between the hard and soft sciences. The foregoing discussion in this chapter and the previous four chapters shows that remaining wedded to the truncated, degenerate conception of causality, concepts, and logic leads to inevitable and irreconcilable conflict between the hard and soft sciences. Weber's reflections on the methodology of social scientific inquiry open up a different possibility. When the categories of causality, concepts, and logic are employed in their full, complete meaning, the deep and intimate relationship between the hard and soft sciences almost forces itself upon the observer or practitioner of scientific activity. The evidence and arguments that Weber has presented to support his position also leaves little room for doubt that his relational conception of science is the more scientific conception, by far.

In the introduction it was noted that this inquiry took Weber's concern about certain developments in modern culture seriously because these developments threatened the health and vitality of the defining ideals of modern Western culture. It was also noted that since Weber's time this crisis has continued to unfold and has been described from different angles by three post-Weber scholars. While Peter Berger explicitly focuses on the crisis in sociology, he notes that this crisis is actually the particular manifestation of a malaise that afflicts not just other social sciences but the human sciences in general. The crisis of coherence in the soft sciences manifests itself in the fact that the tools, methods, and theories employed by the soft sciences do not "stick together" with the social reality that is being studied. One can get a sense of the depth of this crisis in light of four major developments in the post-WWII era: the emergence of the countercultural movement in Western Europe and North America in the 1960s; the rise of East Asian economies in the 1980s; the collapse of the Soviet Union; and the rise of religious fundamentalism (particularly Islamic fundamentalism). Berger notes that: "Each of these developments completely surprised most, if not all sociologists" (Berger, 1992, 13). The analytical tools, research methodologies, and theoretical frameworks used by sociologists could not anticipate the onset of these events. But given the fact that empirical reality in general and social reality in particular is full of surprises and the unexpected, the pundits in the social sciences could be forgiven for not having anticipated these developments. But: "What is more, even after these developments had come sharply into view, sociologists found themselves unable to explain them or to make sense of them within a frame of sociological theory. Given the importance of these developments, the failure of sociology to either predict, or at least to apprehend, them indicates that something is seriously wrong here" (Berger, 1992, 13).

In looking for the cause of the crisis of coherence in the soft sciences, Berger argues that the methodological ailment of the discipline "goes back at the least as far as the 1950's. In a futile and theoretically misguided effort to ape the natural sciences, sociologists developed ever more refined quantitative methods of research" (Berger, 1992, 17). This move had two significant consequences. Firstly, it completely obscured (to the point of negating) the qualitative elements of both empirical and social reality. There is little doubt that quantitative methods shed light on important aspects of social reality.

But other aspects of social reality "are of a character so as to require very different, qualitative approaches. Identification of scientific rigor with quantification has greatly limited the scope of sociology, often to narrowly circumscribed topics that best lend themselves to quantitative methods" (Berger, 1992, 17). Secondly, aping the hard sciences obscured (to the point of negating) the fact that there is a difference between the "rationality" of scientific inquiry and the "rationality" of the empirical reality that is being inquired into: "Sociology, as a science, will necessarily be an exercise in rationality. This is a far cry from assuming that ordinary social action is guided" by rationality as it has been defined by the scientist (Berger, 1992, 17). The one social science that loses sight of this fact most consistently is economics, with the result that "it fails spectacularly, over and over again, to understand, let alone predict, the dynamics of the marketplace" (Berger, 1992, 17). In short: "Yes, sociology is a rational discipline; every empirical science is. But it must not fall into the fatal error of confusing its own rationality with the rationality of the world" (Berger, 1992, 17).

Weber's reflections on causality, concepts, and logic that have been detailed thus far evidence that Weber not only directly speaks to the crisis that is described by Berger but also makes a valuable contribution to the discussion on a number of different levels. Weber's methodology of social scientific inquiry shows how the soft sciences can attain the status of "objective" scientific inquiry without aping the hard sciences. Speaking of Weber's methodological writings (at least one of which predates PESC), Toby Huff notes: "In these methodological writings, virtually all of the logical problems which continue to inspire controversy and debate in the social sciences were given paradigmatic formulation" (Huff, 1984, 73). Huff describes the particular issues that Weber dealt with:

> We find Weber completely rethinking the methodology of the social sciences. To do so, he had to deal with all the fundamental questions about "laws" and "causality," prediction and alleged "irrationality" of human action, the difference between "self-evidence" and "validity," the nature of "understanding" and "interpretation" and of course aspects of the problem of the construction of "ideal-types." (Huff, 1984, 34)

In addition to containing "paradigmatic formulation" of many issues that continue to be debated in late twentieth- and early twenty-first century social theory, Weber's work makes an irenic contribution to these debates. Based on his reading of *The Protestant Ethic and the Spirit of Capitalism*, Stephen Kalberg puts forth the following claim:

> A particular view of the empirical world [emerges], one far distant from the major dichotomies that have structured the perception and understanding of this world among American theorists: history and sociology, conflict and equi-

> librium, society and the individual, culture and economic interest, social order and social change, individual interest and social structure, tradition and modernity and micro and macro levels of analysis. (Kalberg, 1996, 66)

It is not only Weber's view of empirical reality that is "far distant from the major dichotomies that have structured the perception and understanding of this world" among modern scientists. His view of the scientific method that should be used to study this reality is also far distant from the dichotomies and reductionisms that are characteristic of the different theories that are in vogue among not just American theorists but theorists in the Global West. Based on a careful reading of the PESC, Kalberg makes a detailed argument that the methodology of scientific inquiry that is implicit in that work is more advanced than all the dominant methodological positions that emerged in American sociology after WWII. The reason that Weber's position is superior to conflict theory, structural functionalism, modernization theory, comparative-historical sociology, the sociology of culture approach, and rational-choice theory is that it avoids the dichotomies that characterize these other theories. We can take Kalberg's observation further and say that Weber's reflections on the way that the categories of causality, concepts, and logic are employed in the hard and soft sciences respectively, overcomes not only the dichotomies that are characteristic of social theory but of the dominant theories and philosophies of science.

Whereas Berger's analysis of the crisis in the soft sciences help us to better appreciate Weber's continued relevance, Walker Percy's analysis of the crisis in modern science helps us to appreciate the relevance of Peirce. Percy argues that the most pressing crisis in modern culture is the fact that "modern science is itself radically incoherent" (Percy, 1991, 271). In saying that modern science is "radically incoherent," he is referring to something very specific. One manifestation of this incoherence is the fact that while scientific inquiry has exponentially increased our knowledge of the natural cosmos, our understanding of being human and social reality has lost all semblance of coherence: "In the case of the cosmos there is the sense that the areas of ignorance are being steadily eroded by the advance of science. In the case of the sciences of man, however, the incoherence is chronic and seems to be intractable" (Percy, 1991, 273). The other manifestation of incoherence is the fact that the human sciences (or the soft sciences) have no adequate method with which to study the human being in those aspects that make the human being unique among all the phenomena in the universe. For Percy the one thing that makes the human being unique is the fact that the human being is a composite of body and mind. In spite of almost four hundred years of scientific progress since Descartes, we have no rationally coherent account of how the human body is related to the human mind: "I refer to this gap in scientific knowledge as an incoherence, from the Latin *in-cohaerere*, a not-sticking together" (Percy, 1991, 275). This lacuna in scientific knowledge manifests itself in the form of incoherence in the human sciences because no human science has been able to explain how the mind and body "stick together" in human beings. Percy describes the most promising starting point for redressing this particular crisis in these words: "I wish to call your attention to the work of an American scientist who, I believe, laid the groundwork for a coherent science of man, and did so a hundred years ago. Most people have never heard of him. But they will" (Percy, 277). The little-known scientist who is bound to become widely known is none other than Peirce.

It becomes relatively easier to appreciate Percy's valuation of Peirce's work when it is studied in conjunction with Weber's methodology of social scientific inquiry. Social theorists have been trying since the middle of the nineteenth century (if not earlier) to articulate a vision of a "coherent science of man"—or a "scientific science of man." There have been some notable achievements in this regard, as well as major disappointments. Probably the most effective test of Percy's claim that Peirce offers "a coherent science of man" is to see if Peirce's insights are supported by or match those of some major social theorist. In light of the discussion in chapters 3 through 5 we can confidently state the following. When compared with Marx and Durkheim on one side and Geertz and Habermas on the other, it is difficult to identify a social theorist whose methodological reflections come closer to Peirce's phenomenology, semiotics, and logic than Weber. Taken separately, the insights of Weber and Peirce go a long way in redressing the "radically incoherent" condition of the modern sciences. Using different terms and with different points of emphasis, Weber and Peirce offer the logical, methodological, and theoretical capital to bridge the gap between the hard and soft sciences on the one hand and the soft sciences and social reality on the other. As effective as they are by themselves in pointing to a relational, coherent vision of science, the discussion in the first five chapters shows that the insights of Weber and Peirce become even more effective when they are viewed in relational terms.

Having looked at Weber's *Wissenschaftslehre* from the perspective of the methodology and philosophy of science, the discussion will turn to exploring its cultural significance. Weber's sociological inquiry shows that the emergence, development, and evolution of science have been of great significance in human culture at large. But its significance has been significantly amplified in modern Western culture, given the fact that scientific inquiry is one of the most valued cultural activities in the Global West. The next chapter will explore the cultural significance of Weber's relational conception of science—a significance that is brought to the fore when it is contrasted with the impact of the pseudo-scientific conception of science.

NOTES

1. From the above discussion, it is obvious that for both Weber and Peirce the inductive-deductive dyad is woefully inadequate to give an accurate account of the scientific method. They see scientific inquiry to be the result of a triadic relationship between intuitive abductions, deductive generalizations, and inductive testing. Gerald Holton, independently of Weber and Peirce, reached the same conclusion from the perspective of the history of science. He uses the imagery of replacing a two-dimensional plane with three-dimensional space to describe the area in which scientific inquiry occurs. Holton notes that almost all modern philosophies of science describe scientific inquiry in terms of a two-dimensional empirical-analytical plane. After placing the empirical and analytical aspects of scientific inquiry on the x-axis and y-axis of a two-dimensional plane, Holton says "whether they are arbitrary or not, the x-y axes have, since the seventeenth and eighteenth centuries, more and more defined the total allowable content of science and even of sound scholarship" (Holton, 1973, 56). Holton argues that this two- dimensional description of scientific conceptual space is severely limited because it cannot account for the dynamics that are at work during the intense "nascent moment" that is a prelude to novel discoveries. This moment is composed of "explicit or implicit decisions, such as the adoption of certain hypotheses and criteria of preselection that are not at all scientifically 'valid'" (Holton, 1973, 49). At the time that these "explicit or implicit decisions" are being made, they are neither verifiable nor falsifiable according to any criteria known to the scientist—but he makes them nonetheless. Based on close studies of the lab notes and personal reflections of scientists like Kepler, Galileo, Newton, Millikan, and especially Einstein, Holton reaches the conclusion that the practicing scientist is very much aware of the fact that something more than strictly verifiable or falsifiable hypotheses are involved in the actual practice of science. After describing how Newton dealt with hypotheses that were either verifiable or falsifiable, Holton goes on to note that Newton was aware of the fact that a third type of hypotheses/inferences were also present:

> But the fact is that Newton also found one class of hypotheses to be impossible to avoid in his pursuit of natural philosophy—a class that shared with Cartesian hypotheses the characteristic of neither being demonstrable from the phenomena nor following from them by an argument based on induction. . . . The existence, nay, the necessity, at certain stages, of entertaining such unverifiable and unfalsifiable, and yet not quite arbitrary, hypotheses—that is an embarrassing conception which did not and does not fit into a purely positivistically oriented philosophy of science. For the decision whether to entertain such hypotheses is coupled neither to observable facts nor to logical argument. (Holton, 1973, 49)

In order to constructively deal with this class of nonverifiable and nonfalsifiable category of hypotheses, Holton has to go beyond the two dimensional x-y plane and propose a third dimension of scientific conceptual space that has always existed but has remained largely unacknowledged. Holton notes:

> In addition to the empirical or phenomenic (x) dimension and the heuristic-analytic (y) dimension, we can define a third, or z-axis. This third dimension is the dimension of fundamental presuppositions, notions, terms, methodological judgments and decisions—in short, of themata or themes—which are themselves neither directly evolved from, nor resolvable into, objective observation on the one hand, or logical, mathematical, and other formal analytical ratiocination on the other hand. (Holton, 1973, 57)

While the dominant philosophies of science in the twentieth century have remained stuck on the two-dimensional analytical-phenomenic plane, there have been some dissenting voices. But among these dissenting voices, it is difficult to identify a coherent account of the scientific method. One way to get around this issue is to acknowledge the role of subjective, irrational factors in the "context of discovery" and describe the "context of justification" as an objective, rational undertaking. Some of the dissenters who recognize the role of subjective factors in scientific inquiry go so far as to say the presence of irrational, subjective elements in scientific inquiry evidences there is no such thing as a "scientific method." Paul Feyerabend (1984) makes a forceful argument along these lines. The work of Weber and Peirce becomes even more significant against this background because they are able to make a coherent case for a "scientific method," all the while recognizing the role of subjective elements in scientific inquiry.

2. The translation of Weber's description of "causal interpretation" into Peircean terms was done by Peter Ochs.

Chapter Six

The Cultural Significance of Weber's *Wissenschaftslehre*

Having looked at the relational conception of science that is implicit in Weber's *Wissenschaftslehre* and the points on which it differs from the pseudo-scientific conception, the discussion turns to exploring its cultural significance. The point of entry in this discussion is Weber's "disenchantment thesis." Similar to what he says about capitalism, we can say that Weber sees underdeveloped forms of disenchantment in every human civilization that is known to historians. But in its most rational and developed form disenchantment emerges at a particular place and a particular time in history and then spreads to other places at other times in history. Once again following Weber's lead in constructing the investigative question about the causal origins of the "'spirit' of capitalism," we will ask the following question: "While an entire constellation of causal factors led to the rise of disenchantment, which of these factors can be labeled the 'adequate cause'?" In very explicit terms Weber identifies the "progress of science" as the adequate cause of the rise and development of disenchantment. The cultural significance of Weber's *Wissenschaftslehre* starts to come into focus as we refine Weber's answer in light of the points of difference between his relational conception of science and the pseudo-scientific conception that have been detailed in the previous five chapters.

Weber locates the cause of disenchantment in the emergence, development, and evolution of the intellectual sphere of culture, more specifically the processes of intellectualization and rationalization. Intellectualization and rationalization are based on the presupposition that "there are no mysterious incalculable forces that come into play, but rather that one can, in principle, master all things by calculation. This means that the world is disenchanted" (SV, 139). In addition to noting that disenchantment is the ultimate outcome of intellectualization and rationalization, Weber describes the causal factor in this process: "Scientific progress is a fraction, the most important fraction, of the process of intellectualization which we have been undergoing for thousands of years" (SV, 138). Given the fact that the "progress of science" is the "most important fraction" in the process that has led to the disenchantment of the world, we can say (in simple terms) that for Weber "Science is the cause of disenchantment." As we unpacked the seemingly simple statement "Protestantism is the cause of capitalism" to gain access to the relational conception of science that is implicit in Weber's work, we will unpack the Weber's seemingly straightforward "disenchantment thesis" to gain insight into its contemporary cultural significance.

While Weber's sociological inquiry paints an almost fatalistic picture of modern disenchanted culture, if we look at Weber's conception of "progress" carefully an abductive hypothesis emerges that his *Wissenschaftslehre* offers a viable alternative. This hypothesis is in line with an oft-made observation that there is an unresolved tension in Weber's work. This tension and the differing approaches to it in Weber scholarship are summed up by Alan Sica in these words:

> From the beginning of Weber's realization that the Faustian bargain constituting modern Western civilization and industrialization dazzled as well as blinded its beneficiaries . . . his analysis had taken two antipodal forms. Nelson openly admitted, Bendix often obscured and Marcuse joyfully exploited this well-known "tension" in Weber: on one hand, a clear-eyed portraiture of a civilizational condition from which he could not picture a likely escape, yet on the other a strong-voiced denunciation of those "de-humanizing" tendencies which were already well in place 100 years ago. (Sica, 2004, 122)

There is an additional possibility in this tension besides admitting, obscuring, or exploiting it. Keeping the tension in place, Weber's insights give us the ability to not just offer a "strong-voiced denunciation" of the problematic tendencies of disenchantment but also take a principled stand in favor of an alternative—with both the critique and affirmation meeting the highest demands of intellectual integrity.

This chapter will show that while Weber's sociology of culture establishes a causal link between the progress of science and the process of disenchantment, his reflections on the methodology of scientific inquiry (*Wissenschaftslehre*) lays bare the fact that a particular conception of science is the cause of disenchantment, rather than science *per se*. Weber's analysis of the term "progress" shows that the "progress of science" can produce an outcome other than just disenchantment. He describes "progress" as "the quantitative increase and . . . qualitative diversification of the possible modes of response" (MEN, 27) of an individual, society, culture, epoch, etc., in the face of change in their inner and outer environments. He uses the term "progressive differentiation" to describe the increase in the ways that cultural beings can respond to change. Without passing any value judgment on whether this process is a good or bad thing, Weber gives the following description of the outcome of progressive differentiation: "In the sphere of the evaluation of subjective experience, 'progressive differentiation' is to be identified with an increase in 'value' only in the intellectualist sense of an increase in self-awareness or of an increasing capacity for expression and communication" (MEN, 28). Consequently, in addition to leading to disenchantment, progress can lead to increased self-awareness and capacity for self-expression. In sum, the two possible outcomes of the "progress of science" are (a) disenchantment of the world and (b) increased self-awareness and capacity for self-expression.[1] Given Weber's analysis of the differing outcomes of "progress," we can say that both Weber's relational conception of science and the pseudo-scientific conception are the manifestation of the "progress of science." The cultural significance of Weber's *Wissenschaftslehre* become clearly discernible when we focus on the manner in which Weber's relational conception of science contributes to heightened self-awareness on the part of science and increases its capacity for self-expression. While there is no clear and distinct line separating the two, the heightened self-awareness of science that Weber's relation conception produces is evidenced on two different but related issues. Firstly, it allows us to see that disenchantment is not caused by science *per se* but by a very particular conception of science—that is, the pseudo-scientific conception. Secondly, this self-awareness is further heightened by the discovery of "religious" qualities in the cultural praxis of sciences that are laid bare when we look at the constituent elements of Weber's relational conception of science in detail. As a result of this heightened self-awareness, science is able to engage in conversation on venues and across avenues that had been closed down by the pseudo-scientific, disenchanting conception of science—most notably on venues and avenues that cut across the science vs. religion divide.

In more precise terms disenchantment of the world means the death of *Sinn* at the level of theoretical rationalization and perpetual conflict between the different spheres of culture in terms of practical rationalization.[2] The first section of the chapter will show that the truncated, degenerate understanding of the categories of causality, concepts, and logic that are characteristic of the pseudo-scientific conception of science are the very same factors that lead to the death of *Sinn* and inevitable and perpetual conflict between science and nonscience. When we look at the progress of science as it is manifest in Weber's relational conception, we find that Weber's conception contains within it a very different possibility—the opening up of lines of communication and exchange between science and nonscience that the pseudo-scientific conception closes down. The contrasting cultural possibilities between the pseudo-scientific and relational conceptions of science (i.e., inevitable and irreconcilable conflict with nonscience vs. opening up lines of communication) are easily recognized by looking at the starkly different relationship between science and religion that the two contrasting conceptions of science give rise to. This will be an extension of what has been demonstrated in the previous five chapters—Weber's relational conception of science opens up venues and avenues of communication between the hard and soft sciences that the pseudo-scientific conception closes down. In the previous five chapters we saw that Weber's use of the categories of causality, concepts, and logic in their "full meaning" opened up lines of communication between the hard and soft sciences in such a way that it is practically impossible to escape the conclusion that these two types of science are intimately related. The second and third sections of this chapter will expand upon and deepen this point by taking the logic implicit in this conclusion and using it to analyze the relationship between science and the larger culture of which it is a constituent element. This part of the discussion will focus on showing that Weber's analysis lays bare the "religious" dimension of science and the "scientific" dimensions of religion. A key insight (without which Weber's *Wissenschaftslehre* does not make any sense) is the fact that scientific inquiry consists of "subjective" and "irrational" elements. When we look at these "unscientific" elements in science, they are indistinguishable from "religious" elements. The second section will detail this point and show that there is no such thing as scientific inquiry in the absence of faithful affirmation of certain beliefs, no such thing as scientific discovery in the absence of grace, and no such thing as science in the absence of *Sinn*. At the same time, a key insight (without which Weber's sociology of religion and sociology of culture do not make any sense) is the fact that religion consists of "objective" and "rational" elements. When we look at these "irreligious" elements in religion, in the third section, we will find that they are indistinguishable from "scientific" elements. With reference to Weber's sociological and methodological works, the third section will demonstrate that religion is no less worldly, objective, and rational than science.

The chapter will conclude with reflections on the cultural significance of Weber's *Wissenschaftslehre* in the contemporary disenchanted cultural condition. Weber's insights open up the possibility of taking a "heroic stand" in the face of the "fate of our times" on empirically valid and culturally meaningful grounds. This Weberian response to disenchantment is an alternative to the two dominant responses that Weber explicitly recognized: (a) the retreat of religion into the irrational private sphere; and (b) the reduction of science to a meaningless "sporting contest." It is also an alternative to a response that Weber did not see: religious fundamentalism.

THE DISENCHANTING CONCEPTION OF SCIENCE AND THE WEBERIAN ALTERNATIVE

As noted above, Weber began the concluding section of the last public lecture he gave before his death with these words: "The fate of our times is characterized by rationalization and intellectualization and, above all, the 'disenchantment of the world'" (SV, 155). While there is an entire constellation of causes that led to the observed effect of the disenchantment of modern culture, he identifies "scientific progress" as the adequate cause of this process: "Scientific progress is a fraction, the most important fraction, of the process of intellectualization which we have been undergoing for thousands of years" (SV, 138). The final product of scientific rationalization (in intellectual terms) is the fact that it has "created a cosmos of natural causality" (RRW, 355). According to the doctrine of natural causality, all empirical phenomena in the cosmos are the observable effects of impersonal, immutable, and universal "laws of nature." These laws can be precisely quantified and expressed in mathematical form. The doctrine of natural causality has a direct bearing on the relationship of science to religion—it rules out even the theoretical possibility of such a relationship.

Just as the "progress of science" is the cause of disenchantment, the doctrine of natural causality is the cause (i.e., the adequate cause) of the conflict between religion and science. Weber notes that there is an inevitable clash between the doctrine of natural causality and the fundamental presupposition of religion. In spite of many apparent and real differences among religions (up to the point of conflict), religious thought is ultimately based on the presupposition that there is *Sinn* (meaning) in the universe: "At all times and in all places, the need for salvation—consciously cultivated as the substance of religiosity—has resulted from the endeavor of a systematic and practical rationalization of life's realities. . . . On this level, all religions have demanded as a specific presupposition that the course of the world be somehow *meaningful*, at least so far as it touches upon the interests of men" (RRW, 353). The presupposition that the universe and all the events in it are ultimately meaningful has had to be continually justified in the face of an apparently meaningless and chaotic flux of empirical reality—most notably the reality of unjust suffering and undeserved reward among human beings. Religious thought rationalized the claim of meaning in the universe, in the face of empirical evidence to the contrary by formulating the doctrine of "ethical causality." This doctrine reconciled events in empirical reality, in which good things often happened to bad people and bad things often happened to good people, with the claim of *Sinn* (meaning) in the universe. The reconciliation took place by postulating that all of these events are by the will of God and that all human beings will be justly compensated in accordance with a divine standard of retribution "in the end." While the very beginnings of rational, intellectual thought can be traced back to these "religious" roots, there has been a tension between the presupposition of an ethically and meaningfully oriented cosmos and the rational arguments used to support this claim:

> Ethical religiosity has appealed to rational knowledge, which has followed its own autonomous and inner-worldly norms. It has fashioned a cosmos of truths which no longer had anything to do with the systematic postulates of a rational religious ethic; with the result that the world as a cosmos must satisfy the demands of a religious ethic or evince some "meaning." On the contrary rational knowledge has had to reject this claim in principle. The cosmos of natural causality and the postulated cosmos of ethical, compensatory causality have stood in irreconcilable opposition. (RRW, 355)

While a tension has always existed between the religious and intellectual conceptions of causality in the universe, a relationship between rational knowledge and religious knowledge has been historically maintained—but this relationship has come under increasing strain with the passage of time.

At one point in history, the mantle of "rational knowledge" was carried by philosophy, but in more recent times, this mantle has been taken up by science. Consequently, in recent centuries it has been primarily the task of science to defend the doctrine of "natural causality"—and among the most important conditions for defending this doctrine has been to challenge the doctrine of "ethical causality." Weber describes the dynamics characterizing the encounter between religious rationality on the one hand and scientific rationality and the process of disenchantment on the other, in these words:

> The tension between religion and intellectual knowledge definitely comes to the fore wherever rational, empirical knowledge has consistently worked to the disenchantment of the world and its transformation into a causal mechanism. For then science encounters claims of the ethical postulate that the world is a God-ordained and hence somehow *meaningfully* and ethically oriented, cosmos. In principle, the empirical as well as the mathematically oriented view of the world develops refutations of every intellectual approach which in any way asks for a "meaning" of inner-worldly occurrences. (RRW, 350f.)

While the process of disenchantment has been unfolding for thousands of years and there have been many ups, downs, u-turns and cul-de-sacs in the development, in general terms the following is indisputable: "Every increase of rationalism in empirical science increasingly pushes religion from the rational into the irrational realm" (RRW, 351). In spite of this age-old tussle, in the past, religion was able to maintain a modicum of rational integrity. But things have changed dramatically in modern culture because "only today does religion become *the* irrational or anti-rational supra-human power" (RRW, 351). The embrace and triumph of the doctrine of natural causality in the scientific community is the single most important causal factor for consigning religion to the domain of irrationality and expunging the universe of *Sinn* (meaning).

In very direct and no uncertain terms, Weber's analysis of the concept of "causality" demonstrates that the doctrine of natural causality is nothing more the survival of "some variety of modern anthropological occultism" (RK, 194) in modern culture and the modern scientific community. The discussion in chapter 1 showed the logical absurdity of attributing causal agency to any mathematical entity—whether it is a fraction (the fraction 1/6 does not "cause" a particular number to turn up on a throw of the dice) or an equation (F=Ma does not "cause" things to fall to earth or the planets to orbit around the sun). In purely logical terms, attributing the "cause" of the origin of the universe to the "laws of physics" or the origin of life to the "laws of biology" is akin to attributing the origin of capitalism to chemical reactions in the brain and the origin of Buddhism to physiological developments of the skull—in a word, it is pure nonsense. Not only does Weber's *Wissenschaftslehre* provide the logical grounds for rejecting natural causality as occult, pseudo-scientific dogma, it provides the grounds for affirming a scientific alternative—that is, causal imputation. Furthermore, Weber's analysis shows that while the "laws of nature" cannot be accorded causal agency, they are indispensible heuristic devices for conducting the investigation. In order for scientific inquiry to be fully, genuinely scientific, it must raise and address the investigative question: "What is the 'cause' or 'reason' for these heuristic devices to be as they are when they could have been otherwise?" If the character and nature of the "laws of nature" or "human nature" that is attributed causal agency cannot be subject to scientific scrutiny and brought into the domain of rationality, then the final conclusion deserves the label "superstitious dogma" rather than "scientific finding." In addition to demonstrating that the pseudo-scientific conception of causality ultimately produces findings that are (in purely logical terms) practically no better than superstitious dogma, he offers a method of inquiry that produces genuinely scientific results, that is, a method that aspires to causal interpretation of observed phenomena.

In addition to the doctrine of natural causality, the pseudo-scientific conception of science claims that the abstract concept and the controlled experiment are the definitive tools of scientific inquiry. Only those knowledge claims reach the status of being "scientific" in which generic concepts have been used and that have been verified through a lab experiment (or something similar). Weber's analysis of the generic concept has laid bare its shortcomings on at least two different levels. First and foremost, the generic concept is incapable of conceptually capturing the aesthetic and ethical dimensions of empirical reality—especially the aesthetic and ethical elements that the investigator carries with him into scientific investigation. Secondly, the generic concept is incapable of conceptually capturing the unique individual, dynamic variation, and cultural significance. The gross inadequacy of the generic concept is demonstrated by the fact that cultural significance results from the dynamic interaction between values that are held to be "eternal and valid" on the one hand and the unique, concrete reality which is "temporal and in perpetual flux" on the other. Weber describes this process in these words:

> Life with its irrational reality and its store of possible meanings is inexhaustible. The *concrete* form in which value-relevance occurs remains perpetually in flux, ever subject to change in the dimly seen future of human culture. The light which emanates from those highest evaluative ideas always falls on an ever changing finite segment of the vast chaotic stream of events, which flows away through time. (OSS, 111)

Given this dynamic relationship between the concrete real and the eternal "ideal," science must always be prepared to "change its standpoint and its analytical apparatus" (OSS, 112). It is only by being thus prepared that science will keep up with empirical reality. Since the "light of great cultural problems" is always changing and shifting, two things are happening simultaneously. Firstly, different *Sinn* (meaning) is being conferred on an already known finite segment of empirical reality. Secondly, *Sinn* is being conferred on a finite segment of reality that was previously considered to be meaningless. It is as a result of these dynamic processes that new knowledge is being constantly produced. In short, since the scope of "empirical reality" is constantly changing both in quantitative and qualitative terms, science must be in a constant state of readiness to modify its "standpoint and its analytical apparatus" to keep up with the change. From this perspective the meaning and significance of empirical reality can never be exhausted from a scientific point of view: "There are sciences to which eternal youth is granted and the historical disciplines are among them—all those to which the eternally onward flowing stream of culture perpetually brings new problems. At the very heart of their task lies not only the transiency of *all* ideal types *but* also at the same time the inevitability of *new* ones" (OSS, 104).[3] It is only stating the obvious to note that the generic concept is woefully inadequate for this task.

These observations by Weber make us aware of the fact that generic concepts simply cannot capture certain aspects of empirical reality. Instead of admitting their inherent limitations, the claim is made that that which the generic concepts cannot capture and which cannot be expressed in mathematical (or mathematics-like form) is not real. Weber's analysis shows this claim to be a species of the "will-to-believe of naturalistic monism" (OSS, 86). This belief has found a conducive environment for survival in the departments of economics and sociology as much as in the departments of physics, chemistry, biology, and psychology. In the same way that the doctrine of natural causality sets up a clash between religion and science by offering "scientific" grounds for negating the foundational presupposition of religion, the "will-to-believe of naturalistic monism" sets up a clash with religion by positing an unbridgeable gulf between subjective faith and objective rationality. In the previous paragraph we noted Weber's reflections about the relationship between "eternal" values and "changing" reality. The context of these remarks is that they are an exposition of the following point: "We are now at the end of this discussion, the only purpose of which was to trace the course of the hair-line which separates science from faith and to make explicit the *meaning* of the quest for social and economic knowledge" (OSS, 110). Weber makes this observation after spending nearly sixty pages detailing the relationship of "objectivity in the social sciences and social policy" and the issues of causality, concept formation, and logic of scientific inquiry. In simple terms Weber shows us that an "objective" description of these categories is not possible without laying bare and giving an account for the plethora of "subjective" elements that are contained in them. By failing to recognize the fact that "subjective" elements are deeply implicated and intertwined with anything and everything that aspires to "objectivity," the generic concept obscures (or to put it more accurately, negates) the reality of the fact that only a "hair-line . . . separates science from faith." As is the case with the doctrine of natural causality, at the same time that Weber lays bare the pseudo-scientific characteristic of naturalistic monism, he offers a scientific alternative, that is, the ideal-type. The ideal-type makes it possible to study those "subjective" dimensions and characteristics of "objective" scientific inquiry, "objective" empirical reality, and the "objective" scientific study of empirical reality that the generic concept has to categorically reject because of its pseudo-scientific conception of "objectivity."

The clash that natural causality and naturalistic monism set up between religion and science finds its most systematic expression in the confrontation between naturalistic metaphysics and religion. Naturalistic metaphysics sees a direct correlation between the advance of scientific knowledge and the decline of religion. In contrast Weber sees the possibility of the relationship between religion and science being more than a zero-sum game—being fully cognizant of the fact there is grave tension between the two. In a letter to Ferdinand Tönnies, Weber makes the following observation:

> It goes without saying that religions must clash with scientific truth insofar as they assert empirical facts or the causal impact on them of something supernatural. However, when I studied modern Catholic literature in Rome a few years ago, I became convinced how hopeless it is to think that there are any scientific results that this church cannot digest. The steady and slow impact of the practical consequences of our view of nature and history may make these ecclesiastical powers wither away . . . but no anti-clericalism based on "meta-

> physical" naturalism can accomplish this. I could not participate in such anticlericalism. (Schluchter, 1979, 82 fn.)

Then Weber goes on to make the following self-conscious personal evaluation:

> It is true that I am absolutely unmusical in matters religious and that I have neither the need nor the ability to erect any religious edifices within me—that is simply impossible for me and I reject it. But after examining myself carefully I must say that I am neither anti-religious nor irreligious. In this regard too I consider myself a cripple, a stunted man whose fate it is to admit honestly that he must put up with this fate of affairs (so as not to fall for some romantic swindle). I am like a tree stump from which new shoots can sometimes grow, but I must not pretend to be a grown tree. (Schluchter, 1979, 82 fn.)

There are significant implications of a person who is religiously unmusical but not antireligious (and, more significantly, not *irreligious*) when it comes to evaluating the two sides in the religion versus science divide. Weber goes to describe his attitude toward two groups of people: (a) the theologians who are trying to reconcile the findings of science with the teachings of religion; and (b) the scientists trying to replace religion with naturalistic metaphysics. Addressing Tönnies Weber notes:

> From this follows quite a bit: For you a theologian of liberal persuasion (whether Catholic or Protestant) is necessarily most abhorrent as the typical representative of a halfway position; for me he is in human terms infinitely more valuable and interesting . . . than the intellectual (and basically cheap) pharisaism of naturalism, which is intolerably fashionable and in which there is much less life than in the religious position. (Schluchter, 1979, 82 fn.)

The "intellectual pharisaism . . . of naturalism" uses the findings of the natural sciences to construct a "rational" or "objective" or "scientific" metaphysics from which some reason or meaning (or the lack thereof) of empirical reality can be derived. In simple terms, naturalistic metaphysics is the "scientific" attempt to replace religion. In clear, explicit, and no uncertain terms, Weber considers the theologian who tries to bridge the gap between religion and science to be "in human terms infinitely more valuable and interesting" than the (social) scientist who uses the findings of (social) science to construct a metaphysics as a substitute for religion. We need to pause here and identify (with as much precision as possible) the reasons why Weber holds the theologian in such high regard and expresses such disdain for the scientist.

For Weber the scientist who takes it upon himself to construct naturalistic metaphysics to replace religion is practicing a type of "academic prophecy" which should not be confused with "genuine prophecy" (SV, 155). Weber locates the causal origin of academic prophecy (i.e., its adequate cause) in "the need of some modern intellectuals to furnish their souls with, so to speak, guaranteed genuine antiques. In doing so, they happen to remember that religion has belonged among such antiques, and of all things religion is what they do not possess" (SV, 154). To fill the gap that has been caused by loss of religion, these academics "produce surrogates" and "substitutes" of all kinds—one of which is an attempt to produce religion or something religious in the absence of genuine prophecy. In the modern university these attempts take the shape of various theories, philosophies, systems, and the like—all of which rest on some type of scientific or social scientific metaphysics. Weber describes the outcome of these efforts in these words:

> The fate of our times is characterized by rationalization and intellectualization and, above all, by the "disenchantment" of the world. . . . If we attempt to force and to "invent" a monumental style in art, such miserable monstrosities are produced as the many monuments of the last twenty years. If one tries intellectually to construe new religions without a new and genuine prophecy, then, in an inner sense, something similar will result, but with still worse effects. And academic prophecy, finally, will create only fanatical sects but never a genuine community. (SV, 154)

Practically every student in Weber studies has heard of Weber describing disenchantment as the "fate of our times." But in the area of Weber studies specifically and the social sciences more generally, it is rare to find a sustained discussion of Weber's evaluation of the most widely adopted response to disenchantment in the modern university—academic prophecy. One possible reason for this is the fact that the tendencies toward academic prophecy are no less pronounced among Weber scholars and in the soft sciences than they are in the university at large.

In sum, the theologian who attempts to bridge the divide between religion and science is "in human terms infinitely more valuable and interesting" than the academic who attempts to "produce surrogates" for religion via academic prophecy. For Weber the theologian is "infinitely more interesting and valuable" not just because of some aesthetic, ethical, religious, or other such "prejudicial" and "subjective" reasons, but because in the "rational," "objective," and "scientific" position of the secular academic "there is much less life than in the religious position." The attempts of the theologian to bridge the divide between religion and science offer the hope of life for both religion and science in the face of disenchantment—even if it is a hope against hope. In contrast, the attempts to construct a naturalistic metaphysics by the practitioners of pseudo-science (and even more so by the practitioners of pseudo-social science) to replace religion are nothing more than "the evasion of plain duty of intellectual integrity, which sets in if one lacks the courage to clarify one's own ultimate standpoint and rather facilitates this duty by feeble relative judgments" (SV, 155).

The discussion in the previous paragraphs combined with the discussion in the previous chapters teaches us that the clash between religion and science is not the result of the laws of nature/history/psychology, anything inherent in human nature, or anything inherent in the nature of religion or science. The clash is the result of a very particular conception of science that is based on a truncated, degenerate conception of causality, concepts, and logic that produces a pseudo-scientific conception of science. At this point in the discussion we can state Weber's "disenchantment thesis" with greater precision and clarity. While the "progress of science" is indeed the cause of disenchantment, in more precise terms the pseudo-scientific conception of science is the cause of the death of *Sinn* and the conflict between religion and science. Weber rejects this pseudo-scientific, disenchanting conception of science in its totality and without any qualifications. At the same time, Weber offers a "point by point" alternative. The contrast between the disenchanting conception and Weber's relational alternative can be summarized as follows:

Table 6.1.

	The Disenchanting Conception of Science	**Weber's Relational Conception of Science**
Causality	"Anthropological occultism" resting on the doctrine natural causality.	Causal imputation which produces testable hypothesis about motives as the cause of events.
Concept formation	"Will-to-believe of naturalistic monism" resting on the generic concept and controlled experiment.	Ideal-type (containing within it generic and mathematical concepts) and testing hypothesis in lived reality.
Logic of scientific inquiry	"Academic prophecy" that constructs naturalistic metaphysics to replace religion based on inductive and/or deductive philosophies of science and hides the role and impact of the investigator's beliefs and values on the inquiry	Academic integrity that sees scientific inquiry as a dynamic relationship between abductive, deductive, and inductive inferences and lays bare the role and impact of the investigator's beliefs and values on the inquiry.

At this point in the discussion, we see how Weber's analysis has contributed to heightened self-awareness on the part of science by showing that it is not science *per se* that is the cause of disenchantment but a particular conception of science. In the next section we will see that Weber further heightens the self-awareness of science by laying bare the "subjective" elements in science and "objective" elements in religion. The doubly heightened self-awareness of the "progress of science"—as it is manifest in Weber's relational conception of science—shows us that there are deep affinities between science and religion. The recognition of these affinities in their turn opens up venues and avenues of communication between religion and science that the pseudo-scientific, disenchanting conception shuts down.

THE SUBJECTIVE ELEMENTS IN SCIENCE

From the pseudo-scientific perspective, science is a more rational and objective method of attaining knowledge about empirical reality than art and religion because science is free of irrational, prejudicial, and sentimental elements. Whatever their disagreement may be on other points, both the inductivist and deductivist philosophies of science aspire to purge scientific knowledge of all "subjective" characteristics. From Weber's insights we can infer that the claim that scientific inquiry is free of these "subjective" elements and therefore superior to art and religion is the modern incarnation of the Myth of Er—a Noble Lie invented to soothe the fears, placate the fantasies, and affirm the illusions of those who are not strong enough to face the facts and meets the demands of self-conscious and self-critical thought. The facts that Weber's analysis forces every practicing scientist to face is that there is no such thing as science in the absence of three very, very "subjective" elements: (1) A faithful affirmation of certain presuppositions at the beginning of the inquiry; (2) the need of inspiration and grace to conclude the inquiry (in terms of fruitful results); and (3) an acknowledgement of *Sinn* in the universe.

Given the fact that scientific inquiry is a cultural activity that is expressed in the form of a generally agreed upon pattern of action by an identifiable social group, it is inevitable that values and beliefs will be present. To begin with, Weber's reflections make us aware of the fact that there is no such thing as science in the absence of belief and faith. He notes that science is built on certain presuppositions, and the value of science is dependent on their affirmation by a cultural being: "No science is absolutely free from presuppositions and no science can prove its fundamental value to the man who rejects these presuppositions" (SV, 153). At another place: "The means available to our science offer nothing to those persons to whom this truth is of no value. It should be remembered that the belief in the value of scientific truth is the product of certain cultures and is not a product of man's original nature" (OSS, 110). Here Weber is echoing Nietzsche's insight about the precondition that needs to be met before we can have something in empirical reality called "science": "*To make possible for this discipline to begin*, must there not be some prior conviction—even one that is so commanding and unconditional that it sacrifices all other convictions to itself? We see that science also rests on faith; there is simply no science 'without presuppositions'" (Nietzsche, 1974, 280f.). Weber describes one of the presuppositions toward which Nietzsche is pointing by framing the issue in Kantian terms:

> [Kant] took for his point of departure the presupposition: "Scientific truth exists and it is valid," and then asked: "Under which presuppositions of thought is truth possible and meaningful?" The modern aestheticians . . . proceed from the presupposition that "works of art exist," and then ask: "How is their existence meaningful and possible?" (SV, 154)

Weber, no less than Kant, affirms the presupposition that "scientific truth exists and it is valid." In putting forth his own conception of science, Weber is drawing upon the ideas of both Kant and Nietzsche. Even though he benefits from the insights of both, on this particular issue it is clear that he is leaning toward one side:

> Nourished though he was like all German intellectuals in the thought of the greatest of all system builders, Immanuel Kant, and working with the concept that was generative of systemic thought, namely, rationality, in close contact with colleagues like Rickert and Simmel who laid the foundations for systematic sociology, Weber, none the less sided with the hammer of the philosophers, the greatest iconoclast of the modern period. (Albrow, 1987, 164f.)

Weber goes on to note that the type of presupposition on which science rests is hardly any different from the type of presuppositions on which religion, ethics, aesthetics, etc., rest. The presupposition that "scientific truth exists and it is valid" is an "essentially religious and philosophical presupposition" (SV, 154). In order to "*make possible for [the] discipline[of science] to begin*" an "essentially religious and philosophical presupposition" has to be affirmed.

Weber also notes that while the affirmation of this presupposition is a challenge, it is "the least problematic aspect of science" (SV, 143). Beyond the relativity less problematic type of presupposition, there is a presupposition of far greater gravity: "Science further presupposes that what is yielded by scientific work is important in the sense that it is 'worth being known.' In this, obviously, are contained all our problems. For this presupposition cannot be proved by scientific means. It can only be *interpreted* with reference to its ultimate meaning, which we much reject or accept according to our ultimate position towards life" (SV, 143). Given the fact that belief in the value of scientific knowledge is not the product of the laws of nature, anything inherent in human nature, or anything inherent in the nature of science or the nature of values, the question naturally emerges: "On what basis does one accept or reject the value of scientific knowledge?" Weber answers this question directly: "Only on the assumption of belief in the validity of values is the attempt to espouse value-judgment meaningful. However, to *judge* the *validity* of such values is a matter of *faith*" (OSS, 55). Belief in the validity of the value of scientific knowledge is a necessary precondition for a meaningful value-judgment. To go one step further and judge the validity of scientific knowledge (either positively or negatively) is nothing less than "a matter of *faith*."

While scientific inquiry cannot begin in the absence of faith and belief, it cannot end (in the sense of producing useful knowledge) in the absence of grace. Weber notes:

> Inspiration plays no less a role in science than it does in the realm of art. It is a childish notion to think that a mathematician attains any scientifically valuable results by sitting at his desk with a ruler, calculating machines or other mechanical means. The mathematical imagination of a Weierstrass is naturally quite differently oriented in meaning and result than the imagination of an artist, and differs basically in quality. But the psychological processes do not differ. Both are frenzy (in the sense of Platon's "mania") and "inspiration." Now whether we have scientific inspiration depends upon destinies that are hidden from us, and besides upon "gifts." (SV, 136)

The fact that scientific inspiration is dependent much more on "destinies that are hidden from us" and "gifts" than on the abilities of the scientist can be gleaned from the fact that "ideas occur to us when they please, not when it pleases us. . . . Ideas come when we do not expect them, and not when we are brooding and searching at our desks. Yet ideas would certainly not come to mind had we not brooded at our desks and searched for answers with passionate devotion" (SV, 136). Weber makes the same point in a different context: "Politics is made with the head, not with other parts of the body or soul. And yet devotion to politics, if it is not to be frivolous intellectual play but rather genuinely human conduct, can be born and nourished from passion alone" (PV, 115). The pursuit of science, politics, economics, art, or any other cultural activity as a vocation is the result of passionate commitment and passionate commitment alone: "An inner devotion to the task, and that alone, should lift the scientist to the height of dignity of the subject that he pretends to serve. And in this it is not different with the artist" (SV, 137). While passionate devotion rooted in unshakable belief is needed to engage in scientific inquiry, any individual who has attained fruitful results should know that the reason he attained these results has nothing to do with his passion and even less with his ability—the results are the result of grace. Even though he has made this observation with respect to attaining the highest value in the erotic sphere, the exact judgment is applicable to the individual who has attained the highest value in the intellectual sphere: "Rarely does life grant such value in pure form. He to whom it is given may speak of fate's fortune and grace—not of his own 'merit'" (RRW, 350). Ideas "occur to us when they please not when it pleases us" and when they do occur, the one to whom they are occurring should know that they occurred to him because of "fate's good fortune and grace" and not because "of his own 'merit.'"

In addition to making us aware of the role of faith, belief, inspiration, grace, and passionate commitment in the cultural activity called "science," Weber's insights also lay bare the critical role of "meaning" (*Sinn*) in the cultural sciences. As we noted earlier, Weber defines "culture" as "a finite segment of the meaningless infinity of the world process, a segment on which *human beings* confer meaning and significance" (OSS, 81). In light of this definition, it may be possible to define economics without using the word "money" or physics without using the word "atoms," but it is not possible to define the soft sciences without the word "meaning." The progress of science as it manifests itself in the Weber's *Wissenschaftslehre* shows that those parts of the universe that have a bearing on human interests and that interests human are so only because of meaning. Weber goes on to show that meaning is not just constitutive of the universe in which human beings live, it is part and parcel of the "transcendental presupposition of every cultural science"—that presupposition that has to be affirmed before one can formulate a definition of "culture" and subject the human understanding of meaning to scientific scrutiny:

> The transcendental presupposition of every *cultural science* lies not in our finding a certain culture or any "culture" in general to be *valuable* but rather in the fact that we are *cultural beings*, endowed with the capacity and the will to take a deliberate attitude towards the world and lend it significance. Whatever this significance may be, it will lead us to judge certain phenomena of human existence in its light and to respond to them as being (positively or negatively) meaningful. (OSS, 81)

On this point Weber's insights reveal the glaring difference between the disenchanting conception of science and his own relational alternative. As we saw in the previous section, for Weber "all religions have demanded as a specific presupposition that the course of the world be somehow *meaningful*, at least so far as it touches upon the interests of men" (RRW, 353). This presupposition is not just challenged but negated by the hard sciences:

> Who—aside from certain big children who are indeed found in the natural sciences—still believes that the findings of astronomy, biology, physics, or chemistry could teach us anything about the meaning of the world? If there is any such "meaning," along what road could one come upon its track? If these natural sciences lead to anything in this way, they are apt to make the belief

> that there is any such thing as the "meaning" of the universe die out at its very roots. (SV, 142)

In addition to causing the notion of meaning of the universe to "die out at its very roots," the disenchanting conception of science makes the claim that a "scientific," "objective," and "rational" alternative can be constructed. In stark contrast, Weber's relational conception of science remains completely neutral on the issue of the meaning of the universe. But it makes a forceful case that there is meaning in the universe as evidenced by the scientific study of all human beings, at all times, and in all places. The task of scientific inquiry is nothing more and nothing less than objective inquiry into the almost innumerable variety of ways that human beings have invested meaning into the universe, to compare the different ways, and to thereby facilitate a more clear understanding of the meaning that one, oneself, has invested in the universe.

The foregoing discussion shows that Weber's relational conception of science makes us aware of the fact that science is not possible without "irrational" belief, "prejudicial" commitment, and "sentimental" *Sinn.* Not just science, but any nonreligious, worldly, secular activity that aspires to rise above mere "sterile excitation" and be a "genuine passion" (PV, 127) is everywhere and always composed of certain "subjective" elements. Based on the foregoing discussion, we can identify the following set of presuppositions that are common to both religion and science:

1. Meaning is real.
2. Meaning can be had.
3. Meaning is worth having.

While they share these presuppositions, religion and science have qualitatively differing interpretations of these presuppositions. The religious interpretation can be stated as follows:

1. Intuitive/Inspired Vision: The universe is inherently meaningful (according to revealed knowledge).
2. Call to Passionate Devotion: One can and should live a meaningful life (by shaping behavior according to the religious ethic).
3. Faithful Affirmation: A meaningful life is worth living (because this is a necessary precondition for holding onto the hope of attaining redemption from worldly conditions that one finds intolerable).

The scientific interpretation on these presuppositions can be stated as follows:

1. Intuitive/Inspired Vision: Human beings confer meaning on the universe (according to the scientific study of human culture).
2. Call to Passionate Devotion: This meaning can and should be understood (by following the method of scientific inquiry that the soft sciences aspire to do).
3. Faithful Affirmation: This meaning is worth understanding (because it is an expression of the hope of redressing a cultural condition that one finds intolerable).

The fact that religion and science are two differing interpretations of the same set of presuppositions evidences that there is significant affinity between the two. This affinity is very similar to the affinity between the hard and soft sciences that is hinted at in Weber's observation that the difference between the soft and hard sciences "does *not* concern differences in the concept of causality, the significance of concept formation or the kind of conceptual apparatus employed" (RK, 186). Given what Weber's *Wissenschaftslehre* teaches us about the role of inspiration, faith, and meaning in science, we can say that the difference between religion and science "does *not* concern differences in the significance of faith/belief in scientific inquiry, the role of grace in attaining scientific results, and the issue of meaning in the universe." This is not to say that there is no difference between religion and science—on the contrary. Weber gives the criteria for distinguishing between the two: (a) our causal interest in religion differs from our causal interest in science; and (b) the quality of "self-evidence" that we expect from religion is different from the quality that we expect from science.

Before proceeding further we can summarize the discussion in the first two sections as follows. Weber's relational conception of science is an example of the "progress" of science as increase in the self-awareness of science and its capacity for communication. The self-awareness of science has been increased in at least two different ways. Firstly, Weber's insights and analysis show that science *per se* is not the cause of disenchantment but of a very particular conception of science. The particular conception of science that has caused disenchantment is shown to be a truncated, degenerate conception. Secondly, the full, genuine conception of science shows that science is composed of certain sentimental and ethical elements that strongly resemble religious beliefs and values. The increased self-awareness on the part of science that makes it possible to distinguish the disenchanting and relational conceptions in terms of both characteristics and effects, in its turn increases the capacity of science to communicate with nonscience, especially with religion. The same logic and method of analysis that opened up the possibility of seeing the intimate relationship between the hard and soft sciences, has opened up the possibility of seeing affinities between religion and science. This process is furthered when we complement Weber's insights on the "subjective, religious" elements in science with his sociological analysis that lays bare the "objective, worldly" elements in religion. We turn our attention to this part of Weber's work.

THE OBJECTIVE ELEMENTS IN RELIGION

Complementing the claim that science is superior to art and religion because it is free of irrationality, prejudice, and sentiments, the disenchanting conception of science makes the claim that science is superior to art and religion because it values worldly events, critical objectivity, and rationality, whereas religion is otherworldly, rejects critical objectivity, and has no concern with rationality. As was the case with the claim that science is free of irrationality, prejudice, and sentiments, Weber shows the logical absurdity and empirical fallacy of the claim that religion is only otherworldly, lacks critical objectivity, and is not concerned with rationality. Weber's analysis makes us aware of the "worldly, objective" elements of religion, in three different but related senses. Firstly, his sociological inquiry shows just how deeply religion is intertwined with the world and the world with religion. Secondly, he notes that religion has an objective dimension in the form of a pointed critique of religion—a critique to which nonreligious and antireligious thought have added almost nothing. Thirdly, Weber shows that religion is no less rational than any worldly sphere of culture, including science. We will look at each of these points in turn below.

In modern culture religion is considered to be a matter of personal belief about spiritual matters and otherworldly concerns. There can be little doubt that these are important issues with which religion concerns itself. But from the perspective of Weber's sociology, to consider them as the defining concerns and characteristics of religion is a huge mistake. To begin with, Weber notes that from its most primitive beginnings religion has been concerned with the world here and now:

> The most elementary forms of behavior motivated by religious or magical factors are oriented to *this* world. "That it may go well with thee . . . and that thou mayest prolong thy days upon the earth" [Deut. 4: 40] expresses the reason for the performance of actions enjoined by religion or magic. . . . Thus, religious and magical behavior and thinking must not be set apart from the range of everyday purposive conduct, particularly since even the ends of the religious and magical actions are predominantly economic. (ES, 399f.)

Not only are the "most elementary forms" of religious behavior "oriented towards *this* world," the most advanced forms of religious behavior (the conduct of the religious virtuoso) are also purely this-worldly:

> Psychologically considered, man in quest of salvation has been primarily preoccupied by attitudes of the here and now. . . . For the devout the sacred value, first and above all, has been a psychological state in the *here and now*. Primarily this state consists in the emotional attitude *per se*, which was directly called forth by the specifically religious (or magical) act, by methodical asceticism, or by contemplation. (SPWR, 278)

In simple terms, the lay person seeks to attain worldly goods and the virtuoso seeks to attain spiritual goods. At this level there is a clear difference between the two. But there is no difference between them in light of the fact that both of them are seeking their respective ends in the here and now, not in the hereafter. Consequently, "religious and magical behavior and thinking must not be set apart from the range of everyday purposive conduct."

While Weber's sociology of religion shows how deeply rooted religion is in this world, his sociology of culture shows how deeply rooted the different worldly spheres are in religion. He divides the worldly sphere of culture into the economic, political, aesthetic,.erotic, and intellectual spheres. A summary presentation of Weber's sociology of culture is as follows:

Table 6.2.

	Highest Value or Ultimate End	Primary Means of Pursuing the End	Institutional Manifestation
Intellectual Sphere	Knowledge	Science: The abstract concept + experiment	University
Economic Sphere	Profit	Quantification of value in the form of money	Impersonal exchange in marketplace
Political Sphere	Power	Violence	Nation-state claiming monopoly on legitimate use of violence + bureaucracy.
Aesthetic Sphere	Beauty	Free play of imagination	Discovery of new forms in which beauty can be expressed
Erotic Sphere	Pleasure	Sexual union	Lifelong commitment between two, excluding any third.
Religious Sphere	God's Pleasure or Salvation	Living an ethical life	Church, temple, or worldly community of believers.

An inquiry into the origins of each of the different worldly spheres reveals it was originally located in the temple—or has its origins in the religious sphere. At the institutional level, the beginnings of economics, politics, art, and the otherworldly spheres is found to be directly tied to the temple. Weber describes the origins of the worldly spheres in the temple and their subsequent distinction, tensions, and eventually separation from the religious sphere (and from each other) in "Religious Rejections of the World and Their Directions" and "The Social Psychology of World Religions." The way that a particular worldly sphere negotiated its relationship with the religious sphere has proven to be among the most significant factors that have shaped human culture. The following paragraphs will offer an illustrative example of Weber's observation that the origins of the worldly spheres is to be located in the religious sphere by looking at the one part of worldly culture that is as far removed as theoretically possible from religion, that is, antireligious skepticism. This skepticism, formally, belongs to the intellectual sphere and is often thought to be the result of the development of "critical" human rationality and thought. But Weber notes that a study of empirical reality shows that the earliest examples of such thought are not found among (so-called) "free, critical thinkers" but are found in religious scriptures. Not only are the earliest examples of such thought found in scripture, the most compelling arguments that human thought has produced on this topic have hardly gone beyond what is contained in the religious texts. The following observation by Weber captures the depth and breadth of the critique of religion that is found in the sacred scriptures of religious traditions across the globe: "Anti-religious skepticism, *per se*, was represented in China, Egypt, in the Vedas, in post-exilic Jewish literature. In principle, it was just as it is today; almost no new arguments have been added" (RRW, 351).

We can better appreciate Weber's insight on this matter by looking at the Merold Westphal's reading of the Bible. He argues that the "Christian Bible is surely the most anti-religious of all the world's scriptures" and more pointedly it is "a biblical critique of what takes itself to be biblical religion" (Westphal, 1993, 265).For Westphal the depth and breadth of the biblical critique is such that compared to the critique of religion offered by the three iconic modern atheists (i.e., Freud, Marx, and Nietzsche), it is clear that the latter have added "almost no new arguments" to what is already present in the Bible. At first glance this appears to be a fantastical suggestion—to which Westphal responds as follows:

> I believe the final answer to this . . . is found in recognizing the profound parallel between the critique of religion in Marx, Nietzsche, and Freud and the critique of religion found in the Bible. Faith as fraud? Devotion as deception? These are strong charges, but modern atheism is not the first to make them. What about Amos, whose God cannot stand the music offered in his praise (Amos 5: 23)? What about Isaiah (Second or Third) for whom "all our righteous deeds are like a polluted garment" (Isa. 64: 6)? And what about Jesus, who considers the most pious people of his day "whitewashed tombs" (Matt. 23: 27) and the temple run by the chief priests a "den of robbers" (Mark 11: 17)? (Westphal, 1993, 10)

Westphal notes that the critique of religion voiced by James, Paul, and (most importantly) Jesus is the culmination of a tradition that stretches back in history to the ancient Israelite prophets: "We need only recall Jesus' critique of the Pharisees, Paul's critique of works of righteousness, James's critique of cheap grace, and the Old Testament prophetic critiques on which these are based to be reminded that biblical faith has built into it a powerful polemic against certain kinds of religion, even if they are practiced in the name of the one true God" (Westphal, 1993, 11).

The fact that antireligious skepticism is already present in religious texts shows that religion has an "objective" side already present inside of it. As a matter of fact it is difficult to identify a religious scripture in the history of religion in which this "objective" side is not prominent. In addition to being concerned with the world, and the well-being of human beings here and now, the "objective" or "objectivity" is part of religion's self-understanding. In addition to worldly and objective elements, religion also has irreducible rational elements. At a very basic level religion is no less rational than science—if we use a value-neutral and objective definition of "rationality." For Weber the value-neutral definition of rationality is (in simple words) "inner consistency." If judged from this perspective, then the "rationality" attained by modern physics and modern medicine is no different from the rationality of primitive magic (to say nothing of rationality of historical religion). More than any other cultural activity, the disenchanting conception of science links the "reason" for affirming the value of scientific knowledge with "rationality." When asked to give the "reason" she chooses scientific knowledge over nonscientific knowledge (such as magic, religion, art, etc.), the typical scientist will reply that the "reason" is that scientific knowledge is "more rational" than (let's say) magic. Weber's analysis demonstrates that this claim is simply false:

> Magic . . . has been just as systematically "rationalized" as physics. The earliest intentionally rational therapy involved the almost complete rejection of the cure of empirical symptoms by empirically tested herbs and potions in favor of exorcism of (what was thought to be) the "real" (magical, daemonic) cause of the ailment. Formally, it had exactly the same highly rational structure as many of the most important developments in modern therapy. (MEN, 34)

From a formal, logical point of view, it is as nonsensical to claim that modern science is rationally superior to premodern magic as it is to claim that the "laws of nature" are the "cause" of events in empirical reality. If one consistently adheres to the value-neutral, scientific definition of "rationality" as "inner consistency," then it becomes clear that the inner consistency of modern physics, biology, medicine, chemistry, etc., is not quantitatively different from the inner consistency of magic, religion, art, etc. In light of this fact, the typical scientist has to give a reason other than "superior rationality" for affirming the value of science when faced with other equally rational options. If one insists upon using "rationality" as the criterion for distinguishing science from magic, then one will have to make the following concession—the difference between the two rationalities is not one of quantity but one of quality. Quantitatively, both rationalities display the same degree of inner consistency—but they differ significantly in qualitative terms because each of the two rationalities is made possible by the affirmation of two very different sets of "ultimate presuppositions." Complementing his observation that "critical human reason" has not produced any "objective" argument against religion that is not already contained in religious texts, "rational, human reason" has not produced any type of rationality that has not been matched by religious thought. This point can be further detailed by looking at Weber's analysis of theology as a "science" in the sense that physics, biology, and chemistry are sciences.

In the current disenchanted cultural condition, science is considered the most evolved manifestation of intellectual rationality. Weber's analysis shows that religion can equally match the highest form of rational knowledge as it is expressed in science. After talking at length about the definition of science, its impact on occidental history and culture, its place in the university, its constituent elements, and the possibility of pursuing science as a vocation in the face of disenchantment, Weber tackles a very tricky issue: "Which stand does one take towards the factual existence of 'theology' and its claims to be a 'science'? Let us not flinch and evade the answer" (SV, 153). Science is the intellectual rationalization of some particular subject matter—biology is the intellectual rationalization of "life," physics of "matter," economics of "money/wealth," etc. Given an objective and value-neutral definition of science, Weber argues that theology must be labeled as a science because it "represents an *intellectual* rationalization of the possession of sacred value" (SV, 153). There is no difference between theology and the other sciences insofar as it represents the intellectual rationalization of a particular subject matter. What sets it apart from the other sciences is the particular subject matter that it rationalizes, that is, "the possession of a sacred value." The fact that it rationalizes a unique subject matter that is different from the other sciences is no reason to disqualify it as a science, because every science, no more and no less than theology, intellectually rationalizes a unique subject matter. For Weber this makes theology as much of a rational, intellectual enterprise as any other science.

Being a science, theology has the one characteristic that is common to all science; it is based on presuppositions that it itself cannot prove. But there is something that makes theology unique among the sciences. Weber notes that theology goes beyond the "essentially religious and philosophical" presupposition that it shares with science in general. The additional presuppositions are "that 'revelations' are facts relevant for salvation and as such make possible a meaningful conduct of life. Hence, these revelations must be believed in" and "these presuppositions as such lie beyond the limits of 'science'" (SV, 154). This means that "in every 'positive' theology, the devout reaches the point where the Augustinian sentence holds: *credo non quod, sed quia absurdum est*" (SV, 154). In short, while theology is as much of a science as physics, biology, and sociology, it is unique among the sciences because it demands an "intellectual sacrifice" (SV, 155).

In identifying this particular point of difference between theology and the other sciences, Weber is only partially correct. There may have been a time in the past when a scientist could pursue science as a vocation without making an intellectual sacrifice. Even if this was the case in history (and it is highly unlikely that it ever was), a vocational commitment to science in the disenchanted cultural condition is only possible in the aftermath of an intellectual sacrifice. Weber brings to our notice the following "fundamental fact":

> The ultimately possible attitudes towards life are irreconcilable, and hence their struggle can never be brought to a final conclusion. Thus it is necessary to make a decisive choice. Whether, under such conditions, science is a worthwhile "vocation" for somebody, and whether science itself has an objectively valuable "vocation" are again value judgments about which nothing can be said in the lecture-room. To affirm the value of science is a presupposition for teaching here. I personally by my very work answer in the affirmative. (SV, 152)

In simple words, to choose science as a vocation is a value judgment "about which nothing can be said in the lecture-room" but it is a judgment that must be made as a precondition for entering the lecture room. The fact that Weber has chosen science as a vocation and entered the lecture room is clear and compelling evidence of the fact that he has made an intellectual sacrifice—and it differs from the intellectual sacrifice of the religious believer only superficially. It is not only vocational commitment in the intellectual sphere that requires an intellectual sacrifice—a vocational commitment in any of the worldly cultural spheres is only possible in the aftermath of such a sacrifice in a disenchanted cultural condition. Consequently, the one point on which Weber saw theology being different from science (i.e., the intellectual sacrifice) evaporates in the face of progressive disenchantment.

In sum we see that religion is no less concerned with the world, objectivity, and rationality than science. Religion's relationship to the worldly spheres of culture, antireligious objectivity, and rationality show that religion has irreducible "worldly, objective" elements inside of it. The following observation by Alexander shows just how inextricably religion is enmeshed in the "worldly":

> Intellectualization, [Weber] believed, rested upon the most unnatural motivation, led to the most abstracted orientation and inspired the most desiccated organization that the world had ever known. Far from rationality being inherent, it must be understood as the result of a long and complicated evolution of irrational, religious belief. The anti-religious nature of the modern world has a religious basis. (Alexander, 1987, 188)

We concluded the previous section by noting that Weber's insights teach us that the difference between religion and science "does *not* concern differences in the concept of inspiration/grace, the significance of faith/belief, and the issue of meaning in the universe." The present section shows us that the affinity between religion and science goes even deeper by showing that the difference between religion and science "does *not* concern differences in the concept of objectivity, the significance of rationality, and concerns with the world." At the same time that Weber's *Wissenschaftslehre* lays bare the affinity between religion and science, his observations make us aware of the fact that this does not mean that there is no difference between the two. The difference is to be located in the "causal interest" and the "quality of self-evidence" that religion and science respectively pursue. Employing the words that Weber used in a different context, the causal interest of religion in the world is "an interpretation the empirical *validity* of which is problematic" (RK, 191). In contrast, the causal interest of science in the world is "an empirically valid generalization the causal *interpretation* of which is problematic" (RK, 190f.). In short, the depth of relationship between religion and science is indistinguishable from the depth of relationship between the hard and soft sciences (which in its turn hardly differs from the depth of relationship between the natural sciences and physical sciences, which in its turn is practically the same as the depth of relationship between Newtonian physics and quantum physics).

Having looked at the fact that a particular conception of science is the cause of disenchantment (rather than science *per se*) and the fact that Weber's relational conception offers a very different possibility by implying a deep affinity between religion and science, we are in a position to engage in second-order reflections. In the concluding paragraph of *The Protestant Ethic and the Spirit of Capitalism*, Weber notes: "For certain, even with the best will, the modern person seems generally unable to imagine *how* large a significance those components of our consciousness rooted in religious faith have actually had upon culture, national character, and the organization of life" (PESC, 125). One of the reasons that Weber's sociological inquiry is significant is that it identifies the causal role that "religious" factors played in the emergence of a very worldly phenomenon—modern capitalism. Given the mindset of the moderns and their socialization process, the typical modern is "generally unable to imagine *how* large[ly] significan[t]" religion has been in the emergence of modern culture. At the same time, Weber's sociological inquiry also helps us to appreciate the consequences of the complete divorce between the religious sphere and the economic sphere: "Victorious capitalism . . . ever since it came to rest mechanical foundation, no longer needs asceticism as a supporting pillar. Even the rosy temper of asceticism's joyful heir, the Enlightenment, appears finally to be fading" (PESC, 124). Once the relationship between the economic and religious spheres is ruptured, economic activity comes to take on "the character of a sporting contest" and "the idea of an 'obligation to search for and then accept a vocational calling' . . . wanders around in our lives as the ghost of beliefs no longer anchored in the substance of religion (PESC, 124). Under these conditions a unique type of social actors emerges on the stage of history: "Narrow specialists without mind, pleasure-seekers without heart; in its conceit, this nothingness imagines it has climbed to a level of humanity never before achieved" (PESC, 124).

In light of the way that Weber concludes his exposition of the "Protestant ethic" thesis, we can offer the following observations on his "disenchantment thesis." Religion has been of causal significance in the emergence and development of science. Today science, no less than economics, has come to rest on its own mechanical foundations and no longer needs religion as a supporting pillar. When completely freed from its religious roots, we find that two different conceptions of science emerge—the dominant disenchanting conception of science and Weber's relational alternative. At the purely formal and logical level, both are examples of the progress of science. But in substantive terms they are worlds apart.

One way to get a sense of the chasm that separates the disenchanting and relational conceptions is to look at the different types of cultural beings that they give birth to. The disenchanted conception of science necessitates an irreconcilable conflict between religion and science on the one hand and gives birth to the "academic prophet" on the other. "Academic prophecy" is the attempt to construct a naturalistic metaphysics to take the place of religion. As Weber notes repeatedly, science offers something unique and one will seek in vain if one looks for what is unique to science in religion, art, philosophy, ethics, etc. Similarly religion is also unique and it is utter vanity to turn to science, social theory, aesthetic experience, rational coherence, and the like in search for something that only religion can provide. We can go one step further and note that the genus of "conceited nothingness" that has come to flourish in the cultural conditions created by disenchantment is represented by different species in the different worldly cultural spheres. In the intellectual sphere the "academic prophet" is the ideal-typical representative of "conceited nothingness"—the individual who thinks that naturalistic metaphysics or social theory can take the place of religion.

In contrast to the disenchanting conception of science that gives birth to "academic prophets," we find Weber's relational conception of science. In purely technical terms, the significance of this conception is that it shows the deep and intimate relationship between the hard and soft sciences. In cultural terms, Weber's relational conception of science opens up possibilities that the disenchanting conception of science closes down. This alternative possibility is opened by Weber's investigation into the methodology of scientific inquiry which shows that there is deep affinity between religion and science. Just as the disenchanting conception necessitates conflict between religion and science, we saw that Weber's relational conception necessitates deep affinity between the two—the affinities themselves being so deep that they point toward the possibility of relationship. Here Weber is offering a research agenda in which the goal is to see if the affinities between religion and science can be turned into relationships. Undertaking such an attempt in the modern university is foolhardy. To expect it to be taken seriously is sheer madness (especially in the departments of sociology and philosophy). In a word, to follow through on Weber's insights is to attempt the (culturally) impossible.

On this point the following passage can be read as a note of hope and wisdom: "Nonetheless, the possible is often reached by striving to attain the impossible that lies beyond it. Those specific qualities of our culture, which, despite our differences in viewpoint, we all esteem more or less positively, are not the products of the only consistent ethic of '"adaptation" to the possible,' namely, the bureaucratic morality of Confucianism" (MEN, 24). The gravity of the situation and the need to "reach for the impossible" is brought into sharp focus in the following observation:

> Certainly all historical experience confirms the truth—that man would not have attained the possible unless time and again he had reached out for the impossible. But to do that a man must be a leader, and not only a leader but a hero as well, in a very sober sense of the word. And even those who are neither leaders nor heroes must arm themselves with that steadfastness of heart which can brave even the crumbling of all hopes. This is necessary right now, or else men will not be able to attain even that which is possible today. (PV, 128)

An attempt to "attain the impossible" has to be made because if this attempt is not made then "even that which is possible today" will become unattainable tomorrow. In sum, the cultural significance of Weber's *Wissenschaftslehre* is that it offers the theoretical possibility of an alternative to the "fate of times." At the same time it contains insights and pointers that are compelling and powerful enough to give an individual the courage to "reach out for the impossible." Without any pretension of being a hero or leader but with sober self-awareness, an attempt will be made in the next chapter to reach for the impossible. This attempt will seek to transform the religion vs. science divide that is the necessary outcome of the disenchanting conception of science into a relationship based on Weber's insights about the deep affinities between religion and science. In pursuing this task we will bring two parts of the present discussion into relationship. The discussion in the first five chapters showed that a relational vision of the hard and soft sciences is implicit in Weber's work. This relational vision was made explicit using Peirce's phenomenology, semiotics, and logic. The discussion in this chapter put forth evidence that a close reading of Weber's work shows that there are deep affinities between religion and science. In the next chapter we will bring these two parts of the discussion into relationship by looking at the Weberian affinities between religion and science from the perspective of Peirce's pragmaticism in our attempt "to attain the impossible" that lies beyond "the fate of our times."

NOTES

1. Weber's conception of "progress" is far more detailed and nuanced than presented here. One of the places where he discusses it at length is (MEN, 26–36). For a general overview, see (Koshul, 2005, 129–137).

2. This point is detailed in chapter 1 (Koshul, 2005).

3. Here Weber is describing the continuous, evolutionary process by which irrational, chaotic, and meaningless empirical reality is brought into the sphere of rationality, culture, and meaningfulness. This is a continuous, dynamic, evolutionary process that can never end, even theoretically, as long as there are cultural beings. The terms in which Weber describes the "eternal youth" that has been granted to certain sciences resonates deeply with John Deely's description of the process of semiosis in which inanimate things (genuine Secondness) are turned into objects (degenerate Secondness) and then into signs (Thirdness). Deely notes: "In semiosis, a sign brings something not semiotic into the semiotic realm: it makes of a thing an object signified, a significate, an element reticulated in the net of experience, leading in turn to further signifieds, many of which are often new" (Deely, 1994, 167). Using Deely's language to express Weber's position, we can say that culture is a semiotic activity in which meaningless things and objects are brought into the "net of experience," turned into signs (i.e., icons, indices, or symbols) and thereby brought into the domain of *Sinn*. Deely's interpretation of Peirce's semiotics has the potential to further clarify and deepen the link between Weber and Peirce.

Chapter Seven

Weber, Peirce, and a Relational Vision of Religion and Science

The goal of this chapter is to see if Peirce's insights can help us to turn the deep affinity between religion and science that is implicit in Weber's work into a mutually enriching relationship. In the attempt to transform the Weberian affinity into a relationship, we will turn to philosophy—philosophy in the full, genuine sense of the word, not in its truncated, degenerate form. The truncated, degenerate form of philosophy limits itself in a number of ways. Firstly, only a select few individuals from among all of humanity have the innate abilities to engage in its pursuits. Secondly, its subject matter, that is, "wisdom," is something that is not only set apart from but also worlds away from the everyday mundane reality that is experienced by the masses. This is philosophy that defines itself as "the love of wisdom" and then goes on to set up conditions that puts "wisdom" beyond the reach of all women, peasants, shopkeepers, laborers, artisans, and the like—who collectively constitute more than 95 percent of humanity. In contrast to this limiting (and limited) conception of philosophy, Peirce offers a fuller and more expansive conception: "The kind of philosophy which interests me and must, I think, interest everybody is that philosophy, which uses the most rational methods it can devise, for finding out the little that can as yet be found out about the universe of mind and matter from those observations which every person can make in every hour of his waking life" (1.126). For Peirce philosophy is not so much the pursuit of "wisdom" as something which is set apart from everyday mundane reality as it is the pursuit of that knowledge which can be attained by the disciplined study of everyday mundane reality. The contrast between Peirce's conception of philosophy and the degenerate conception becomes even more pronounced in light of the fact that "most philosophers set up a pretension of knowing all there is to know—a pretension calculated to disgust anybody who is at home in any real science" (1.128). The primary reason why this pretension is sure to disgust scientists is because the self-proclaimed lover of wisdom thinks that philosophy is a self-sufficient endeavor and wisdom can be had without any regard for what science has to say about the universe of mind and matter. Peirce takes strong exception to this attitude: "All we have to do to turn our backs upon all such truly vicious conduct, and we shall find ourselves enjoying the advantages of having an almost virgin soil to till, where a given amount of really scientific work will bring in an extraordinary harvest" (1.128). While scientific knowledge about the universe of mind and matter is important in Peirce's conception of philosophy, it is outstripped in importance only by the practice of scientific inquiry. Given the dismissive and derisive attitude of the truncated, degenerate conception of philosophy towards science we can label it the pseudo-scientific conception of philosophy. In contrast, given the central role that scientific knowledge and practice play in Peirce's fuller and more expansive conception of philosophy we will call it by the name that he gives it "scientific philosophy."

The difference between scientific and pseudo-scientific philosophy is brought to the fore when we consider their respective objects of inquiry. The object of inquiry of pseudo-scientific philosophy is something called "wisdom." The line of inquiry here revolves around issues of what is wisdom, how can it be attained, and how do we know that we have attained wisdom. The object of inquiry in Peirce's scientific philosophy is something radically different. Scientific philosophy begins with the observation that "the special sciences are obliged to take for granted a number of most important propositions, because their ways of working afford no means of bringing these propositions to the test" (1.129). Whether we are talking about physics, biology, and chemistry, or mathematics, mechanics, and meteorology, or sociology, economics, and history—in short, all the hard, soft, and hard-soft sciences "assume general principles that cannot be proved or disproved by their ordinary methods of work" (1.129). In other words, the special sciences "always rest up on metaphysics" (1.129) and its validity cannot be tested using the tools of not just the particular science in question but the tools of all of the special sciences taken together. First principles and metaphysics are such an integral part of the special sciences that Peirce argues: "Find a scientific man who proposes to get along without any metaphysics . . . and you have found one whose doctrines are thoroughly vitiated by the crude and uncriticized metaphysics with which they are packed" (1.129). Given the fact that every science rests on assumptions, postulates, first principles, and metaphysics that none of the special sciences can possibly test, the question naturally emerges, "On what grounds are we to repose confidence in these untested suppositions?" Peirce notes: "The philosopher alone is equipped with the facilities for examining such [general principles on which the special sciences rest but which cannot be proved or disproved by their ordinary methods of work] and for determining the degree to which confidence may safely be reposed in them" (1.129).

The contrast between scientific philosophy and pseudo-scientific philosophy is clear enough. It would be worthwhile to go on a brief tangent and offer some reflections on the relationship between the truncated, degenerate conception of causality, concepts, and logic that is characteristic of the pseudo-scientific conception of science on the one hand and pseudo-scientific philosophy on the other. Since its origins in ancient Greece, philosophy has been defined as the "love of wisdom" and the claim has been made throughout the centuries that philosophers are capable of rising above their personal and cultural prejudices and arriving at universal truths by using purely objective methods and the rational intellect alone. That there is something profoundly wrong with this claim is shown by the fact that when we explore the historical "origins of philosophies" (Dewey, 1920, 5) we find that "philosophy originated not out of intellectual material, but out of social and emotional material" (Dewey, 1920, 25). With the benefit of historical hindsight we can see that "philosophy did not develop in an unbiased way from an open and unprejudiced origin" (Dewey, 1920, 18) because from the very beginning the task of philosophy and philosophers has been "to justify on rational grounds the spirit, though not the form, of accepted beliefs and traditional customs" (Dewey, 1920, 18). "Subjective" elements such as emotions, ethical, social, and cultural concerns have been at least as important in shaping philosophy as "objective" elements such as rationality, objectivity, and scientific criteria. Dewey notes that without taking the "subjective" elements that are specific to Greek culture into account, there is no way to make sense of Greek philosophy:

> After all, Plato and Aristotle reflected the meaning of Greek tradition and habit, so that their writings remain, with the writings of the great dramatists, the best introduction of a student into the inner most ideals and aspirations of distinctively Greek life. Without Greek religion, Greek art, Greek civic life, their philosophy would have been impossible; while the effects of that science upon which the philosophers most prided themselves turns out to have been superficial and negligible. (Dewey, 1920, 19)

If Western philosophy is indeed an expansive footnote on Plato (or on ancient Greek philosophy more generally) then we can say that the "'spirit' of Western philosophy" cannot be separated from the "ancient Greek ethic."

While Dewey records the impact of different aspects of Greek culture on the development of Greek philosophy, he largely sidesteps its implications for our understanding of the term "philosophy." Robert Jenson's reflections bring us face to face with the implications, which are as unsettling as they are far-reaching. With reference to the nearly two-millennia-old debate about the difference between "theology" and "philosophy," Jenson notes:

> We usually refer to the work of Greece's theologians with their own name for it, "philosophy."
>
> We have thereupon been led to think this must be a different *kind* of intellectual activity than theology, to which theology perhaps may appeal for foundational purposes or against which theology must perhaps defend itself. But this is a historical illusion; Greek philosophy was simply the theology of the historically particular Olympian-Parmenidean religion, later shared with the wider Mediterranean cultic world. (Jenson, 1997, 10f.)

From this perspective, Greek philosophy is nothing more than the theology of ancient Olympian-Parmendian religion. This means that the philosophies of Heraclitus, Parmenides, Plato, Aristotle, and Plotinus are examples of "pagan theology," and the theologies of Origen, Athanasius, Augustine, Anselm, and Aquinas are examples of "Christian philosophy." Combining Jenson's linguistic analysis with Dewey's historical reflections, we can see that the "Olympian-Parmendian ethic" is the cause (i.e., the adequate cause) of the rise of the "'spirit' of Western philosophy."

When we look at the unique characteristics of the truncated, degenerate notions of causality, concepts, and logic that underpin the pseudo-scientific conception of science, it is difficult to escape the conclusion that their origins are located in beliefs that confer legitimacy on the "Olympian-Parmendian ethic" and confer meaning (or lack thereof) of the *Lebensführung* that is the "'spirit' of Western philosophy." A summary account would be as follows:

Table 7.1.

Modern Pseudo-Scientific Conceptions of Causality, Concepts, and Logic	Aristotle's Interpretation of Olympian-Parmendian Beliefs	Plato's Interpretation of Olympian-Parmendian Beliefs
Natural Causality	Unmoved Mover	Demiurge
Generic Concept	Universal Essence	Ideal Form
Inductive-Deductive Naturalistic Metaphysics	Necessarily Existent Being	Immutable Realm of Ideas

From this perspective, two points come into focus. Firstly, there is a great deal of affinity between the core elements of the pseudo-scientific conception of science on the one hand and the leading principles (i.e., exactitude, certainty, and universality) of pseudo-scientific philosophy on the other. Secondly, Weber's *Wissenschaftslehre* represents a radical break with classical Greek philosophy (as well as its ossified remains that manifest themselves in the form of positivist, intuitionist, deductivist, and inductivist philosophies of science).[1] His conception of adequate cause isolates a particular factor from an entire constellation of factors as the cause of a phenomenon rather than the Cause, thereby laying bare the perspective of the investigative question of inquirer. His conception of the ideal-type shows that empirical reality is composed of unique individuals, qualities, beliefs, and values, rather than Universals. And his conception of causal interpretation helps us to identify the reason of particular events being as they are when could have been otherwise rather than Reason and sets the stage to begin inquiry into the *Sinn* of empirical reality (especially inquiry into the *Sinn* of the investigation itself).

If Nietzsche is right then the "spirit" of Western philosophy is composed of two elements; a "game" and a "shadow." In his *Philosophy in the Tragic Age of the Greeks*, Nietzsche notes that the pre-Socratic philosophers reached the conclusion that the end goal of all the motion, change, and flux that is observed in material reality is, in the worlds of Heraclitus, nothing more than "a game" (Nietzsche, 1998, 112). Offering his own judgment he notes: "This seems to me to have been the final solution, the ultimate answer, that ever hovered on the lips of the Greeks" (Nietzsche, 1998, 112). While the goal of observed motion is the manifestation of a divine "sporting contest," material reality for post-Socratic philosopher is, in the words of Plato, nothing more than "a shadow." The "spirit" of Greek philosophy which treats material reality with utter contempt and disdain permeates, through and through, the conceptions of causality, concepts, and metaphysics on which the pseudo-scientific conception of science is built. In offering an alternative conception of causality, concepts, and causal interpretation, Weber opens up the possibility of looking at events in material reality as meaningful and significant occurrences. At the same time his relational conception of science requires that we treat material reality as something far more than a "shadow," "illusion," or a "mirage"—we treat it as "real," as "real" anything else in Reality.

The foregoing distinction between scientific philosophy and pseudo-scientific philosophy can be summed up thus. Scientific philosophy is not some vague, hazy "love of wisdom" but a particular branch of science whose "function is to trace the uncritical assumptions of human thought to their hiding places" (Iqbal, 2012, 1). Given the fact that human knowledge claims in general, and scientific knowledge claims in particular, are always and everywhere underpinned by "uncritical assumptions," philosophy becomes the most crucial of sciences in helping us to determine "the degree to which confidence may safely be reposed in" our knowledge claims—especially our scientific knowledge claims. Among us, the scientists (whether they are practitioners of the hard sciences or the sciences) are in special need of philosophical training and literacy. This is due to the following fact:

> Every man becomes more or less imbued with philosophical opinions, without being clearly aware of it. Some of these, it is true, may be right opinions; if he is a quite uneducated man, they doubtless will be so. . . . The more a man is educated in other branches, but not trained in philosophy, the more certain it is that two-thirds of his stock of half-conscious philosophical opinions will be utterly wrong, and will completely blind him to the truth, which he will gradually become unable so much as to conceive. (1.134)

The educated man and practicing scientist are in far greater need of scientific philosophy than the uneducated, illiterate peasant. This is due to the fact that the intuitive instincts and common sense of the illiterate peasant is sound and healthy, and more often than not a healthy instinct leads to true philosophical opinions. In contrast, the training of the scientist and educated man has so damaged their intuitive faculties and common sense that 66 percent of their philosophical opinions are "utterly wrong" and they are "completely blind" and incapable of even "conceiving," let alone recognizing, philosophical truth. The end goal of scientific philosophy is not "wisdom" but looking at the "reasoning processes which" have "worked well" in light of experience and then inquiring "into the conditions of their working well" (1.165). The one reasoning process that has worked particularly well in the light of experience of the past few hundred years comes under the general heading of the "scientific method." When we inquire "into the conditions" or the "reasons" and "causes" of the scientific method working well, we will find that Peirce's scientific philosophy almost necessarily leads to the transformation of the Weberian affinities between religion and science into relationships.

The first section of this chapter will focus on the Weberian insight that religion and science are two different interpretations of the same presuppositions: (1) Meaning is real, (2) Meaning can be had, and (3) Meaning is worth having. The last complete essay that Peirce wrote before he died was titled "A Neglected Argument for the Reality of God." This essay shows that Peirce's philosophy of religion and philosophy of science are based on the exact same phenomenology, semiotics, and logic. The relationship between the religion and science is so intimate that the objective truth of scientific knowledge is dependent on the reality of God—and vice versa. The second section of the chapter will focus on Weber's insight that objective, scientific investigation of empirical reality is not possible in the absence of subjective sentiments and values. In his essay titled "The Doctrine of Chances," Peirce explicitly identifies the sentiments that must be presupposed for scientific logic to function. These sentiments are faith, hope, and charity. As every student of Western intellectual history knows, not only are these sentiments (in the full, genuine sense) unknown in ancient Greek philosophy, they are not even theoretically conceivable. In other words, objective scientific inquiry is not possible in the absence of specifically religious sentiments. The third section will focus on Peirce's description of the objective elements in religion by looking at his essay "A Religion of Science." This essay describes the possibility of a vision of religion that is so confident in itself that it can submit its conception of *Sinn* to the rigors of scientific investigation and tests of empirical validity. When we look at Peirce's conception of "a religion of science" we can derive a conception of a "science of religion"—this is a science that is so confident in itself that it can place its conception of empirical validity in the orbit of religious *Sinn*. The discussion will conclude by looking at Peirce's essay titled "Evolutionary Love." This essay helps us to see that while the pseudo-scientific, disenchanting conception of science and Weber's relational conception of science are products of the same evolutionary process, the former is an example of truncated, degenerate evolution while the latter is an example of evolution in its full, genuine meaning. At the conclusion of the discussion after the first five chapters, we were able to see that the truncated, degenerate, pseudo-scientific conception of science almost necessarily closes down the possibility of relationships between the hard and soft sciences that the full, genuine, scientific conception of science opens up. In the same vein, when we compare the degenerate, pseudo-scientific conception of evolution with the genuine, scientific conception, we will find that the latter opens up cultural horizons and possibilities that the former necessarily closes down.

THE REALITY OF GOD AND THE LOGIC OF SCIENCE

The last complete essay that Peirce wrote before he died in 1914 is titled "A Neglected Argument for the Reality of God." The title of this work indicates that it is primarily concerned with matters of God, religion, and belief. But there is much more to the essay. Michael Raposa notes that the essay "despite the metaphysical and cosmological themes that pervade it, turns essentially on certain claims made about the logic of inquiry" (Raposa,1989, 32). Both the structure and content of the essay show that Peirce's argument for the Reality of God is inseparable from the pragmaticist logic of scientific inquiry. Besides this particular essay, Raposa notes that Peirce's ideas on religion and religious belief shows that his "reflections on religious topics are continuous with, even integral to, his work in semiotic, metaphysics and the normative sciences" (Raposa, 1989, 3). If Raposa is correct, then we cannot separate Peirce's philosophy of science from his philosophy of religion, and the "Neglected Argument" presents compelling evidence that this is indeed the case.

Peirce uses the term "Neglected Argument" in two different (but related) senses. First, it refers to "a nest of three arguments for the Reality of God" (6.486). Second, it refers to the second of the three arguments in the nest. We will look at each of the three arguments in the nest individually—taken together they encapsulate Peirce's NA. Peirce calls the first of the threefold nested arguments the Humble Argument (HA). The argument begins with the activity of "Pure Play": "Now, Play, we all know, is a lively exercise of one's powers. Pure Play has no rules, except this very law of liberty. It bloweth where it listeth. It has no purpose, unless recreation" (6.458). Pure Play is primarily concerned with "some wonder in one of the universes, or some connection between two of the three, with speculation concerning its cause" (6.458). Here the term "universe" has to be understood in the context of Peirce's phenomenology. The categories of Firstness, Secondness and Thirdness (or consciousness, fact, and reason) refer to three different "universes." Pure Play is focusing on "some connection" between two of the three universes—especially concerning the origin of cause of the connection. Peirce notes: "It is this last kind—I will call it 'Musement' on the whole—that I particularly recommend, because it will in time flower into the N.A." (6.458). This Musement should be allowed to unfold for forty to fifty minutes without any interruption and eventually it will yield its fruits: "In the Pure Play of Musement the idea of God's Reality will sooner or later to be found an attractive fancy, which the Muser will develop in various ways" (6.465). Continued reflection on the ideas of God's Reality will make the idea not only increasingly attractive but also increasingly relevant: "The more he ponders it, the more it will find response in every part of his mind, for its beauty, for its supplying an ideal of life and for its thoroughly satisfactory explanation of his whole threefold environment" (4.465).

After the hypothesis of God has emerged in the mind, the attention of the Muser is fixed on the hypothesis. Among other things, the Muser becomes curious about its origin —perhaps more curious about this than any other aspect of the hypothesis. At this stage the Humble Argument gives way to the Neglected Argument. The Neglected Argument locates the origin of the hypothesis in the Muser's intuitive instincts and also shows that: "Far from being a vicious or superstitious ingredient, it is simply the natural precipitate of meditation upon the origin of the Three Universes" (6.487). Peirce notes that to deny the reality of instincts (and more importantly its results) is to deny that human beings belong to the world of nature. There is practically no doubt regarding the reality of instincts working in the animal kingdom. For Peirce, instinct is also at work in the human world (more specifically in the world of human thought). In (6.476) Peirce draws an analogy to illustrate how ridiculous it is to claim that human instinct does not exist or that it is not reliable. To the degree that a human being is a part of the natural world, instinct is as much a part of the human being as it is of any bird or animal—and as reliable. During the course of scientific inquiry it is very often the case that the inquirer is faced with a choice of two or more hypotheses that can be used to explain the same set of facts. Preferring one hypothesis over the other(s) is almost always a case of instinctual preference: "Yes; it must be confessed that *if we knew* that the impulse to prefer one hypothesis to another really were analogous to the instincts of birds and wasps, it would be foolish not to give it play, within the bounds of reason; especially since we must entertain some hypothesis, or else forego all further knowledge than that which we have already gained by that very means" (6.476). Stated in simpler terms, belief in God in the aftermath of meditation on natural phenomena originates in the same instinctive impulse that helps us to select one among a number of likely hypotheses during the course of scientific inquiry.

After naming and describing the Humble and Neglected parts of the NA, Peirce does not give a name to the third part of the argument. But he does give a clear description:

> The third argument, enclosing and defending the other two, consists in the development of those principles of logic according to which the humble argument is the first stage of a scientific inquiry into the origin of the three Universes, but of an inquiry which produces, not merely scientific belief, which is always provisional, but also a living, practical belief, logically justified in crossing the Rubicon with all the freightage of eternity. (6.485)

Whatever the name of the third part of the NA, it has to perform two tasks—serve as a conclusion to the NA and show that the first part of the NA (the Humble Argument) produces not just "scientific belief" but also a "living, practical belief" that is "logically justified." From this it is obvious that Peirce sees significant similarities between the logic of scientific discovery and the last part of the NA. But instead of going on to describe the third part of the NA, he starts to describe the logic of scientific inquiry in detail. We will come back to this apparent tangent after completing the NA.

The Muser began by contemplating some phenomenon in the universe(s) and the hypothesis of God's Reality emerged. Reflecting on the origin of the hypothesis, the Muser discovers that the origin is her intuitive instinct. At this stage the contemplation will leave the universe(s) and the origins of the hypothesis behind and focus on the reality that the hypothesis suggests—the reality of God. Continued contemplation will widen and deepen the Muser's proximity to God. As the intensity of the proximity increases, the character of the Muser begins to be affected by this relationship. Peirce uses an analogy to explain this point: "Just as long acquaintance with a man may deeply influence one's whole manner of character" (6.502), so will an individual's acquaintance with God become the determining factor in shaping his own character. As a matter of fact, this is the closest that we can come to having a precise description of God—a relationship that improves the character of the individual. Speaking of what the term "God" means for a pragmaticist, Peirce notes: "So if contemplation and study of the physico-psychical universe can imbue a man with principles of conduct analogous to the influence of a great man's works or conversation, then that analogue of a mind—for it is impossible to say that *any* human attribute is *literally* applicable—is what he means by 'God'" (6.502). The degree to which the claimant modifies his habits to bring them in line with the knowledge claim, to that degree the knowledge claim is valid. Stated in more precise, technical terms: "The true meaning of any product of the intellect lies in whatever unitary determination it would impart to practical conduct under any and every conceivable circumstance, supposing such conduct to be guided by reflexion carried to an ultimate limit" (6.490).

In light of what Peirce has said in other places, but using the terms he uses in the first part of this essay, we can call the third part of the NA the Pragmaticist Argument. Taken together, the Humble, Neglected, and Pragmaticist Arguments make up the NA. At this stage we begin to see what Raposa meant when he said that the NA "despite the metaphysical and cosmological themes that pervade it, turns essentially on certain claims made about the logic of inquiry" (Raposa, 1989, 32). Ochs describes the third part of the NA as "a methodeutic of pragmaticism" (Ochs, 1998, 235) because Peirce acknowledges that "the presentation of this argument would require . . . a strict proof of the correctness of the maxim of Pragmaticism" (6.485). Leaving the proof for later, Peirce begins a new subsection of the essay titled "The Three Stages of Inquiry" right after finishing the NA. After informing the reader that the remainder of the essay is composed of three subsections, the introductory paragraph of this subsection describes the relationship between the subject matter of the first subsection and the subject matter of the previous section (the NA) in these words: "The first shall give the headings of the different steps of every well-conducted and complete inquiry, without noticing possible divergencies from the norm. I shall have to mention some steps which have nothing to do with the Neglected Argument in order to show that they add no jot nor tittle to the truth which is invariably brought just as the Neglected Argument brings it" (6.468). In Weber's terms, Peirce is constructing an ideal-type of the conceptual apparatus that is employed in "every well-conducted and complete [scientific] inquiry"—of which the "Neglected Argument for the Reality of God" is an example (and perhaps the very best example). Peirce notes that in constructing the ideal-type he has to go into details that were missing in the NA. But practically speaking, these details add "no jot nor tittle to the truth which is invariably brought just as the Neglected Argument brings it." With these introductory words, Peirce goes on to construct an ideal-type of the three stages of scientific inquiry (i.e., Abduction, Deduction, and Induction.)

The intimate relationship between the logic and method of scientific inquiry on the one hand and the hypothesis of the reality of God on the other is illustrated by the way in which Peirce addresses possible criticisms of his argument. He acknowledges that there are various philosophies of science and various theologies that can critique the pragmaticist conception of the reality of God. But he rejects these critiques on purely logical grounds. To begin with, this conception can be critiqued on the basis of philosophical skepticism. Peirce argues that if we accept the presuppositions of skepticism we will have to reject the validity of science as well: "If we cannot in some measure understand God's mind, *all* science . . . must be a snare" (8.168). While the beginning of the following passage explicitly refers to skepticism in general, the end of the passage leaves no doubt as to the specific type of skepticism that is being critiqued:

> Scepticism, in the sense of doubt of the validity of elementary ideas—which is really a proposal to turn an idea out of court and permit no inquiry into its applicability—is doubly condemned by the fundamental principle of scientific method—condemned first as obstructing inquiry and condemned second because it is treating some other than a statistical ratio as a thing to be argued about. No: as to God, open your eyes—and your heart, which is also a perceptive organ—and you see him. (6.493)

For Peirce philosophical skepticism with respect to the reality of God is a position that "the scientific method utterly condemns" (6.493).

Apophatic theology is different from philosophical skepticism in that it affirms the reality of God but then becomes skeptical regarding any positive knowledge claims about God. From this perspective the only reasonable things we can say about God are in purely negative terms—what God is not. Any positive statements about the attributes of God are considered unreasonable by definition. Peirce summarizes the apophatic position in these words:

> Various great theologians explain that one cannot attribute *reason* to God, nor perception (which always involves an element of surprise and of learning what one did not know) and, in short, that his "mind" is necessarily so unlike ours, that some—though wrongly—high in the church say that it is only negatively, as being entirely different from everything else, that we can attach any meaning to the Name. (6.502)

While skeptical philosophy has to be rejected for violating the foundational principle of the scientific method, negative theology has to be rejected in the face of the actual results (or fruits) of scientific inquiry. Peirce argues that we have to reject the claim that "we cannot attach any meaning to the Name" because "the discoveries of science, their enabling us to *predict* what will be the course of nature, is proof conclusive that, though we cannot think any thought of God's, we can catch a fragment of His Thought, as it were" (6.502).

Peirce is aware of the implications of his conception of God for the logic of scientific inquiry. His conception of God is such that the answer to the question "whether all physical science is merely the figment—the arbitrary figment—of the students of nature" (6.503) is directly dependent on the answer to the question whether God is Real or not. Put in simple terms, if God is Real then scientific inquiry yields true knowledge; if God is not Real then scientific knowledge yields illusions. Peirce notes: "The hypothesis of God's Reality is logically not so isolated a conclusion as it may seem. On the contrary, it is connected so with a theory of the nature of thinking that if this be proved so is that" (6.491). Among other things, the "Neglected Argument for the Reality of God" is an argument in which one cannot separate what is commonly called "religious belief" from what is commonly called the "scientific method." The intimate relationship between the two can be illustrated by looking at the source of the pragmatic maxim—the third part of the NA, which also describes "certain parts of the work of scientific discovery" (6.457). In technical terms, the maxim is: "Consider what effects, that might conceivably have practical bearings, we conceive the object of our conception to have. Then, our conception of these effects is the whole of our conception of the object" (5.402). Peirce inserts the following comment in the footnote to (5.402): "It has been said to be a sceptical and materialistic principle. But it is only an application of the sole principle of logic which was recommended by Jesus; 'Ye may know them by their fruits,' and it is very intimately allied with the ideas of the gospel" (5.258). While different pragmatists may differ on other issues, all of them affirm the validity of the pragmatic maxim. The following observation describes the maxim in simpler terms, along with its genealogy:

> All pragmatists will further agree that their method of ascertaining the meanings of words and concepts is no other than that experimental method by which all the successful sciences . . . have reached the degrees of certainty that are severally proper to them today; this experimental method being itself nothing but a particular application of an older logical rule, "By their fruits ye shall know them." (5.465)

Here Peirce is alluding to the Gospels, more specifically to Matthew 7:15–20:

> Beware of false prophets, who come to you in sheep's clothing but inwardly are ravenous wolves. You will know them by their fruits. Are grapes gathered from thorns or figs from thistles? In the same way, every good tree bears good fruit, but the bad tree bears bad fruit. A good tree cannot bear bad fruit, nor can a bad tree bear good fruit. Every tree that does not bear good fruit is cut down and thrown into the fire. Thus you will know them by their fruits. (Matthew 7:15–20)

Peirce's Neglected Argument shows that the logic of scientific inquiry cannot be separated from the Reality of God. They are based on the same phenomenology, semiotics, and logic. Peirce explicitly states that the truth of one is dependent on the truth of the other. In Weberian terms we have seen that religion and science are two differing interpretations of the same presuppositions: (1) Meaning is Real, (2) Meaning can be had, and (3) Meaning is worth having. The religious interpretation of these presuppositions produces "meaning" (*Sinn*) whose validity has to be tested. The scientific interpretation of these presuppositions produces "validity" whose meaning has to be affirmed. Peirce shows us in order for religion to reach full maturation (have its meaning validated) it is in need of scientific truth. Conversely, in order for science to become complete (have its validity be meaningful), it needs religion. This point will be detailed in the third section when we look at Peirce's essay "A Religion of Science." Before looking at the essay we will see how Peirce deepens the relationship between religion and science by laying bare the "religious" sentiments that are at the root of scientific reasoning.

RELIGIOUS SENTIMENTS AND SCIENTIFIC LOGIC

In chapter 5 we saw that Peirce's reflections show that thought is composed of ideas, connections between ideas, and the habit of connecting ideas. Looking at the habit of connection of ideas, we found that it is not possible in the absence of the sentiment of belief. We also noted that for Peirce besides beliefs other sentiments are constituent elements of scientific thought. We will describe this point in more detail. Peirce argues that there cannot be any scientific reasoning in the absence of something like religious sentiments. As we have seen, abduction, deduction, and induction are three different types of inferences that collectively constitute the scientific method. Peirce analyzes the reasons why we can invest confidence in the results of inferential reasoning. He argues that a certain "sentiment [is] presupposed in reasoning" (2.655)—in the absence of this sentiment, logically speaking, there can be no inferential reasoning. He details this argument in two of his papers, titled "The Doctrine of Chances" (2.645–668) and "Grounds of Validity of the Laws of Logic" (5.318–357). In simple terms the sentiment can be described as "the social impulse." A close look at this sentiment reveals that it is supported by two other sentiments. In total, Peirce identifies three sentiments that are presupposed in scientific reasoning:

> It may seem strange that I should put forward three sentiments, namely, interest in an indefinite community, recognition of the possibility of this interest being made supreme and hope in the unlimited continuance of intellectual activity, as indispensable requirements of logic. Yet, when we consider that logic depends on a mere struggle to escape doubt, which, as it terminates in action, must begin in emotion and that, furthermore, the only cause of our planting ourselves on reason is that other methods of escaping doubt fail on

> account of the social impulse, why should we wonder to find social sentiment presupposed in reasoning? As for the other two sentiments which I find necessary, they are so only as supports and accessories of that. (2.655)

We will look at the different types of inferences and the role of the different types of sentiments in logically validating them.

Peirce notes that there is a difference between deductive inference on the one hand and inductive and abductive inferences on the other hand. The validity of a deductive inference (necessary reasoning) is based on the presupposition that true premises necessarily lead to true conclusions. The validity of inductive and abductive inferences (or probable reasoning) is based on the presupposition that there is similarity and continuity between the known and the unknown. For Peirce, it is practically impossible to doubt the validity of the presupposition on which deductive logic is based. The reasoning of deductive logic is as follows: "That if one sign denotes generally everything denoted by a second and this second denotes generally everything denoted by a third, then the first denotes generally everything denoted by the third, is not doubted by anybody who distinctly apprehends the meaning of these words" (5.320). The basic presupposition on which deductive logic is based is that "some things are necessarily true." The syllogism that Peirce offers as an example of deductive logic is also an example of that which is necessarily true—if the two premises are true, the conclusion is necessarily true. Peirce examines the different skeptical arguments that question the validity of deductive logic (5.327–332). He goes on to show the reason why these skeptical arguments fail. In addition to skeptical arguments that attempt to undermine this validity, a variety of sophisms have been put forward to undermine the adequacy of deductive inferences. Peirce takes these up in turn (5.333–340) and demonstrates that the sophisms themselves are based on internal contradictions. The upshot of this part of Peirce's discussion is that up till this point no argument has been put forward that has been able to undermine the presupposition on which the validity of deductive inference is based. In the end, he recognizes that even though the skeptical and sophist positions are invalid, a degree of difficulty still remains in demonstrating the validity of the grounds of the validity of deductive logic. Peirce notes that the "difficulty of showing how the law of deductive reasoning is true depends upon our inability to conceive of its not being true" (5.341).

After establishing the grounds for the validity of deductive inferences in a roundabout way, Peirce notes that establishing the validity of inductive and abductive inferences presents a very different set of problems. In an induction we make an inference about the whole based on our observations of a part. We project the results obtained from an observed portion of reality to the unobserved portion of reality—assuming that the same holds true for the unobserved part as is true of the observed part. In an abduction we make an inference about the state of affairs in an unexperienced future based on our knowledge of state of affairs in the experienced past. In both cases we assume that there is continuity and relationship among the observed and experienced part of reality on one hand and the unobserved and unexperienced part of reality on the other. Human experience shows that human beings make correct inductive and abductive inferences not infrequently. For Peirce it is nothing short of wondrous and magical that human beings can attain knowledge of that which they have not experienced—but at the same time it also gives rise to a profound difficulty in establishing the validity of the grounds of inductive and abductive logic:

> In the case of probable reasoning the difficulty is of quite another kind; here, where we see precisely what the procedure is, we wonder how such a process can have any validity at all. How magical it is that by examining a part of a class we can know what is true of the whole of the class and by study of the past can know the future; in short, that we can know what we have not experienced! (5.341)

The fact that we can have knowledge about that which we have not directly experienced suggests that human beings have certain capacities beyond the merely physical and intellectual that make it possible for them to know that which they have not physically or intellectually experienced. Taken collectively, Peirce calls these capacities "intellectual intuition":

> Is not this an intellectual intuition! Is it not that besides ordinary experience which is dependent on there being a certain physical connection between our organs and the thing experienced, there is a second avenue of truth dependent only on there being a certain intellectual connection between our previous knowledge and what we learn in that way? Yes, this is true. Man has this faculty, just as opium has a somnific virtue. (5.341)

Peirce goes on to ask "how is the existence of this faculty accounted for?" (5.341).

In providing an account for the existence of "intellectual intuition," Peirce draws out attention to the conditions that must be met in order for a probable inference to be true:

> All probable inference, whether induction or hypothesis [abduction], is inference from the parts to the whole. It is essentially the same, therefore, as statistical inference. Out of a bag of black and white beans I take a few handfuls and from this sample I can judge approximately the proportions of black and white in the whole. This is identical with induction. Now we know upon what the validity of this inference depends. It depends upon the fact that in the long run, any one bean would be taken out as often as any other. (5.349)

All probable inferences are based on the assumption of a "long run"—or indefinite repetition of an activity. The "long run" in turn suggests that while we may not be confident in a particular inference that we have made about a particular event, we are nonetheless confident in the method that has led up to this particular inference. The reason for this confidence rests on the fact that the habit of mind that induction and abduction generate has produced accurate (and therefore desirable) results in the past. This means that moving from knowledge of the known to knowledge of the unknown is not possible in the absence of two elements: (a) particular inferences about particular phenomena, and (b) a method of drawing inferences that we are confident in—for Peirce it is faith in the latter that makes it possible to invest confidence in the former:

> Upon our theory of reality and of logic, it can be shown that no inference of any individual can be thoroughly logical without certain determinations of his mind which do not concern any one inference immediately; for we have seen that that mode of inference which alone can teach us anything, or carry us at all beyond what was implied in our premises—in fact, does not give us to know any more than we knew before; only, we know that, by faithfully adhering to that mode of inference, we shall, on the whole, approximate to the truth. (5.354)

In sum, "faithfully adhering to that mode of inference" that has proven to be beneficial among a community of inquirers is the first sentiment that is needed to affirm the validity of the laws of scientific logic.

When we look at the issue more closely we see that we need to add something more to the discussion. The community to which we belong is not limited to our immediate historical and geographical location—this community extends indefinitely into the past and also across geographical boundaries of all types. It is difficult to identify a group of people at any point in time known to historians who have not made some type of contribution to the body of knowledge that we label "scientific." Not only does this community cut across borders and into the misty past—it reaches out into the indefinite future. In order for us to logically affirm the validity of our inductive and abductive inferences today, we must entertain the hope that the experience of those who come after us (more or less) confirms the validity of this method—and this "tomorrow" has to be extended into the indefinite future:

> Now, there exist no reasons . . . for thinking that the human race, or any intellectual race, will exist forever. On the other hand, there can be no reason against it; and, fortunately, as the whole requirement is that we should have certain sentiments, there is nothing in the facts to forbid our having a *hope*, or calm and cheerful wish, that the community may last beyond any assignable date. (2.654)

For Peirce the role of the sentiment of hope is so central to inferential reasoning that even an atheist cannot properly reason in its absence: "This infinite hope which we all have (for even the atheist will constantly betray his calm expectation that what is Best will come about) is something so august and momentous, that all reasoning in reference to it is a trifling impertinence" (5.357).

As important as faith and hope are, they are only corollaries to the all-important sentiment of charity. The importance of this ultimate sentiment can be illustrated by looking at the importance of community. The habit of drawing inferences that must be faithfully affirmed in order for reasoning to be logically valid is not the habit of an isolated, atomistic individual—it is the habit of an individual who is part of a larger community. Similarly, the hope in the future is not merely the hope of one's own future but the hope for the future of the community. Given the fact that both faith and hope are significantly tied to the community, the individual must put the interest of the community ahead of her (or his) individual interests. If an individual is selfish, then logically speaking the inference drawn by the individual is illogical. Peirce describes the importance of this sentiment in quite dramatic terms:

> It seems to me that we are driven to this, that logicality inexorably requires that our interests shall *not* be limited. They must not stop at our own fate, but must embrace the whole community. This community, again, must not be limited, but must extend to all races of beings with whom we can come into immediate or mediate intellectual relation. It must reach, however vaguely, beyond this geological epoch, beyond all bounds. He who would not sacrifice his own soul to save the whole world, is, as it seems to me, illogical in all his inferences, collectively. Logic is rooted in the social principle. (2.654)

To put the interests of others before one's own interests is often described by the adjective "charitable"—the all-important sentiment needed to establish the validity of the laws of logic is charity.

The section began by describing the sentiments of scientific reasoning in technical jargon. Using Peirce's language, the sentiments identified as the indispensible requirements for logic were described as "interest in an indefinite community, recognition of the possibility of this interest being made supreme and hope in the unlimited continuance of intellectual activity." During the course of the discussion, the three sentiments were translated into simpler terms—faith, hope, and charity. This translation is also warranted by Peirce's writing. Reflecting on his own technical description of the three sentiments, Peirce notes:

> It interests me to notice that these three sentiments seem to be pretty much the same as that famous trio of Charity, Faith and Hope, which, in the estimation

> of St. Paul, are the finest and greatest of spiritual gifts. Neither Old nor New Testament is a textbook of the logic of science, but the latter is certainly the highest existing authority in regard to the dispositions of heart which a man ought to have. (2.655)

We noted as early as chapter 3 that, for Weber, anywhere we find patterned action in empirical reality, it is clear evidence that aesthetic and ethical elements are also present. For Weber beliefs, patterned action, and values are the three basic constituent components of empirical reality. Peirce's description of the grounds of validity of the laws of logic lays bare the "subjective" beliefs and values that always and everywhere accompany the "objectively" observed pattern of action called "the scientific method." Combining the terminology of Weber and Peirce, we can say that the "idea" of science in the mind of the practicing scientist is not possible if the sentiments of faith, hope, and charity are not present in the heart of the scientist.

FROM SENTIMENT AND LOGIC TO RELIGION AND SCIENCE

We begin by looking at Peirce's definition of science and religion respectively. For Peirce science is far more than just "systematized knowledge" as the dictionary defines it. The shortcoming in this definition is that "mere knowledge, though it be systematized, may be a dead memory; while by science we all habitually mean a living and growing body of truth" (6.428). He goes on to note that knowledge may not even be "necessary to science" because a scientist may conduct an investigation and reach a false conclusion but the inquiry will still be labeled "scientific" because the method that was followed was the scientific method. Consequently, "that which constitutes science . . . is not so much correct conclusions, as it is a correct method" (6.428). Furthermore, the method itself "did not spring out of the brain of a beginner: it was a historic achievement and a scientific achievement" (6.428). In order to answer the question "What is science?" we must identify the origins of the historic and scientific achievements that we call the scientific method. Peirce describes this origin in these words:

> So that not even this method ought to be regarded as essential to the beginnings of science. That which is essential, however, is the scientific spirit, which is determined not to rest satisfied with existing opinions, but to press on to the real truth of nature. To science once enthroned in this sense, among any people, science in every other sense is heir apparent. (6.428)

Since the discussion at hand is Peirce's understanding of "science," it is worth recalling Peirce's description of a scientific intelligence as being "an intelligence capable of learning by experience" (2.227). It is also the case that the enterprise of science is not a solitary quest: It can only be carried out by a community of inquirers. Stated in summary form, for Peirce science is:

- the "spirit" of pressing on "to the real truth" irrespective of established opinions
- being open to changing one's ideas in the face of new experiences
- being part of a community of inquirers

Peirce goes on to give his description of religion. To begin with, religion is a sentiment that can be found in each and every individual:

> And what is religion? In each individual it is a sort of sentiment, or obscure perception, a deep recognition of a something in the circumambient All, which, if he strives to express it, will clothe itself in forms more or less extravagant, more or less accidental, but ever acknowledging the first and last, the {A} and {Ö}, as well as a relation to that Absolute of the individual's self, as a relative being. (6.429)

While it is a sentiment in an individual, it not merely a personal sentiment because: "Religion cannot reside in its totality in a single individual. Like every species of reality, it is essentially a social, a public affair. It is the idea of a whole church, welding all its members together in one organic, systemic perception of the Glory of the Highest" (6.429). Contrary to popular conception, the characteristic of "growth" is among the defining features of religion. Having already described religion as one of the species of "ideas," Peirce notes that it is "an idea having a growth from generation to generation and claiming a supremacy in the determination of all conduct, private and public" (6.429). In summary, religion is:

- a sentiment that perceives the reality of an alpha and omega to which the individual's self is related
- a growing idea that claims supremacy in determining all public and private conduct
- a reality that is embodied in a community

Given the defining features of science and religion identified above, the question naturally arises: Is there any relationship between the two? A superficial view of the issue suggests that the answer is "No." Peirce acknowledges that this conclusion is justified in light of the history of science and the history of religion, respectively. It is obvious that "those who are animated by the spirit of science are hurrying forward" (6.430), ceaselessly searching for new horizons, new possibilities, and new realities. In contrast, the adherents of religion are marked by a conservative spirit that cherishes the memories of past events and deceased individuals. With the passage of history the "destiny [of religion] is to wilt and fade. The vital sentiment that gave it birth loses gradually its pristine purity and strength, till some new creed treads it down" (6.430). From this perspective the contrast between science and religion is as stark as it is clear: "Science, to specialists, may seem to have little or nothing to say that directly concerns religion; but it certainly encourages a philosophy which, if in no other respect, is at any rate opposed to the prevalent tendency of religion, in being animated by a progressive spirit" (6.431). Additionally science is primarily concerned with investigating the facts of material reality. Given this focus, scientists often display a "tendency to pooh-pooh at things unseen" (6.431). Religion, for its part, considers the unseen domain of reality as more real than the visible domain.

Under these circumstances it appears that there is no real possibility of a meaningful relationship between science and religion. The foregoing discussion suggests that this possibility is precluded for two basic reasons. First, science cannot compromise its progressive spirit and remain science, while religion cannot abandon its conservative spirit and remain religion. Second, science cannot make anything besides visible material reality its primary concern, while religion cannot make anything other than unseen reality its primary concern. Peirce is aware of the factors that have created a vast chasm between science and religion—a chasm that was the defining characteristic of the (non)relationship between the two in Peirce's own day. He surveys this chasm with the critical eye of a scientist as well as the faithful hope of a believer and foresees the possibility of the chasm being bridged. Peirce posits that this bridge will be built by individuals who have "a state of mind" that he calls "a religion of science" (6.433). He begins his description of "a religion of science" by first highlighting what it is *not*. It is not "a religion to which science or the scientific spirit has itself given birth; for religion, in the proper sense of the term, can arise from nothing but the religious sensibility" (6.433). Speaking of the "religion of science" as an attitude, Peirce notes: "This attitude, be it observed, is one which religion will assume not at the dictate of science, still less by way of a compromise, but simply and solely out of a bolder confidence in herself and in her own destiny" (6.433). This means that the religion of science will not be an artificially concocted phenomenon produced by "academic prophecy" (to use Weber's term) that appeals to the finding of science and even less to the findings of social science.

Arising from the natural sentiment that is the source of all genuine religious sensibility, the religion of science will produce believers who are able to reconcile the progressive spirit of science with the conservative spirit of religion in their persons. Peirce describes these individuals in the following terms:

> While adhering to the essence of religion and so far as possible to the church, which is all but essential, say, penessential, to it, he will cast aside that relig-

> ious timidity that is forever prompting the church to recoil from the paths into which the Governor of history is leading the minds of men, a cowardice that has stood through the ages as the landmark and limit of her little faith and will gladly go forward, sure that truth is not split into two warring doctrines and that any change that knowledge can work in his faith can only affect its expression, but not the deep mystery expressed. (6.432)

For these believers every new discovery of science becomes a cause for celebration because it enriches them in the one thing that is most dear to them—their religious life. Their religious life is enriched by new scientific discoveries because these discoveries provide new signs, indicies, and symbols that allow them to describe "the circumambient All . . . the first and the last, {A} and {Ö}." Never losing sight of the fact that the description of the circumambient All "will clothe itself in forms more or less extravagant, more or less accidental," it is nevertheless the case that the past, present, and future discoveries of science render that description less extravagant and less accidental. In Weberian terms, the "religion of science" describes an attitude where believers are not afraid of letting their interpretation of the *Sinn* of the universe be subject to tests of logical coherence and empirical validity in light of the findings of science.

The fact that Peirce can envision a religion of science emerging in the future evidences that he sees science and religion sharing relational affinities that have gone unnoticed. These affinities can be categorized under two broad headings: (1) the respective origins of science and religion; and (2) the respective methods of acquiring scientific knowledge and attaining religious belief. When we look at the respective origins of science and religion, it becomes clear that both of them are rooted in human sentiment. Peirce explicitly describes religion as being a "sentiment" (6.429). As we have seen in the Neglected Argument, Peirce identifies something that is very similar to sentiment, that is, "instincts," as the origin of science. He notes that all the "triumphs of science" fall into one of two categories: "They either consist in physical—that is, ultimately, dynamical—explanations of phenomena, or else in explaining things on the basis of our common sense knowledge of human nature" (6.500). Keeping in mind that "dynamics is nothing but an elaboration of common sense; its experiments are mere imaginary experiments," it logically follows that "it all comes down to common sense in these two branches" (6.500). When we look closely at common sense in these two branches it becomes apparent that these are two different manifestations of human instincts: "The one is founded on those instincts about physical forces that are required for the feeding impulsion and the other upon those instincts about our fellows that are required for the satisfaction of the reproductive impulse" (6.500). The conclusion that logically follows is that "all science is nothing but an outgrowth from these two instincts" (6.500). It is worth noting that this passage is taken from Peirce's essay titled "Answers to Questions Concerning My Belief in God." He concludes the passage with this observation: "You will see that all I have been saying is not preparatory to any argument for the reality of God. It is intended as an apology for resting the belief upon instinct as the very bedrock on which all reasoning must be built" (6.500). To the degree that scientific reasoning is a species of reasoning, it too rests on the "bedrock" of instinct. Peirce's discussion of "the grounds of validity of the laws of logic" can be seen as a detailed exposition of this point.

In the same way that Weber's scientific conception of science allows us to see the deep and intimate relationship between the hard and soft sciences that the pseudo-scientific conception of science hides, Peirce's conception of scientific philosophy allows us to see the deep and intimate relationship between religion and science that pseudo-scientific philosophy hides. From the perspective of Peirce's pragmaticism, scientific truth cannot be separated from belief in God—and God in a very particular sense. The term "God" has nothing to do with some pseudo-scientific, impersonal concept variously called the Prime Mover, Monad, Transcendental Ego, Universal Mind, and other such artificial concoctions. For Peirce God is "the first and last, the {A} and {Ö}" (6.429). In philosophical terms, Peirce describes the relationship between the Alpha, the Omega, and the universe in these terms: "The starting-point of the universe, God the Creator, is the Absolute First; the terminus of the universe, God completely revealed, is the Absolute Second; every state of the universe at a measurable point of time is the third" (1.362). In simple terms, for Peirce this is the God of the Hebrew Bible and the Gospels (and to the degree that the Qur'an is a part of this tradition, also the God of the Qur'an). This is the God who created the universe and is the source of "genuine prophecy" (in Weber's terms) and spoke to Abraham, Moses, David, Jesus, and Muhammad (as well as other Noble Prophets).

There is an intimate relationship between this particular conception of God and the foundational presupposition of scientific philosophy: "To believe in a god at all, is not that to believe that man's reason is allied to the originating principle of the universe?" (2.24). For Peirce the relationship between reason and God is deeper and more intimate than a mere hypothetical affirmation. When speaking of the spirit in which philosophy should be studied, Peirce notes: "The spirit in which, as it seems to me, philosophy ought to be studied is the spirit in which every branch of science ought to be studied; namely, the spirit of joy in learning ourselves and in making others acquainted with the glories of God" (1.127). In a slightly different context and in slightly different words, Peirce makes the same point—the purpose of the theoretical sciences is "simply and solely the knowledge of God's truth" (1.239). From the perspective of his scientific philosophy, it is clear that the truth and glories of God cannot be separated from the universe that God has created—the very same universe that is the object of scientific investigation. For Peirce it is not possible to become "acquainted with the glories of God" and "God's truth" without becoming acquainted with the universe because the universe is "a great symbol of God's purpose, working out its conclusions in living realities" (5.119). Being a "great symbol of God's purpose" is another way of saying that the "Universe [is] precisely an argument" (5.119) for the reality of God. Peirce goes on to describe the character of the argument that the universe represents in more detail: "The Universe as an argument is necessarily a great work of art, a great poem—for every fine argument is a poem and a symphony—just as every true poem is a sound argument" (5.119). After noting that poetry is "one sort of generalization of sentiment," Peirce notes that the "complete generalization of sentiment is religion, which is poetry, but completed poetry" (1.676). In describing the Universe that science studies as "a great poem" which is making a "sound argument" and religion as "completed poetry," Peirce is pointing the way toward the emergence of "a science of religion"—that science which embraces the teachings of religion as its own not because of external compulsion but because of an inner calling. This would be a science that sees in religion something that it lacks in itself but needs in order to be fully, genuinely itself rather than a truncated, degenerate form of what it could and should be. It is at this point in the discussion that we explore the vocation of science (not to be confused with science as a vocation) in the modern, disenchanted cultural condition.

Weber notes that every scientist not only knows that "what he has accomplished will be antiquated in ten, twenty, fifty years," the scientist "cannot work without hoping that others will advance further than" he has (SV, 138). Given the fact that the inner ethos of science is tied to the continuous progress of scientific knowledge, the following situation and accompanying question emerges:

> In principle, this progress goes on *ad infinitum*. And with this we come to inquire into the *meaning* of science. For, after all, it is not self-evident that something subordinate to such a law is sensible and meaningful in itself. Why does one engage in doing something that in reality never comes, and never can come, to an end? (SV, 138)

Putting the same issue in slightly different terms, the following question and situation emerges:

> Has "progress" such a recognizable meaning that goes beyond the technical so that to serve it is a meaningful vocation? The question must be raised. But this is no longer merely the question of man's calling for science, hence, the problem of what science as a vocation means to its devoted disciples. To raise

> this question is to ask for the vocation of science within the total life of humanity. (SV, 140)

On one level the answer to the question "What is the meaning of science?" is simple and straightforward. As noted in chapter 6, quoting and agreeing with Tolstoy, Weber notes: "Science is meaningless because it gives no answer to our question, the only question important for us: 'What shall we do and how shall we live?'" (SV, 143). While science cannot give an answer to this most important of questions, it can render a uniquely valuable service to the individual who asks the question. If the teaching and learning of science is done with integrity, science "can force the individual, or at least . . . help him, to give *an account of the ultimate meaning of his own conduct*" (SV, 152). At the current stage in the discussion, we can rephrase the question in the following terms: "What is *the ultimate meaning* of that conduct that comes under the heading 'scientific inquiry'?" We have already seen in chapters 2 through 4 that "meaning" emerges only as the result of a relationship—a relationship between certain facts and certain values. Consequently, in order for the vocation of science to have any meaning, science must be in relationship with values. Keep in mind that it is not the business of science to determine the validity of values; the values with which science will have to establish a relationship will have to be derived from nonscience (or from the spheres of culture other than the intellectual sphere). Furthermore, since scientific inquiry belongs to the intellectual sphere which is a part of the "rational" spheres of culture (along with economics and politics), in order for meaning and significance to emerge in the "rational" spheres of culture, they will have to be brought into relationship with the "irrational" spheres of culture (i.e., the erotic and aesthetic). And furthermore, in order for any meaning and significance to emerge in the "rational" and "irrational" spheres of culture taken together, they will have to be brought into relationship with the religious sphere. The meaning and significance of science in modern culture (or the vocation of science) will be dependent on the type of relationship that it has with nonscientific elements of culture. Weber's methodological reflections, taken together with Peirce's logical inquiry, show that in the absence of such a relationship all claims of meaning and significance are empirically invalid. A summary of Peirce's insights brings into sharp focus just what is included in the phrase "nonscience." In the three foregoing sections we saw the following:

- Peirce's "Neglected Argument for the Reality of God" shows that the scientific method that is used to arrive at scientific truth is intimately related to the method of religious reflection that generates belief in God that the truth of scientific knowledge is logically dependent on the truth of belief in God (and vice versa).

- Peirce's description of the sentiments that are precondition for scientific rationality shows that they are the very sentiments that religion considers to be "the finest and greatest of spiritual gifts."
- Peirce's analysis of the logical relationship between the Universe that is studied by science and the belief in God shows that the former is nothing more and nothing less than a poetic argument for the latter.

In light of Peirce's findings, we can rephrase Weber's question about the value of science in the total life of humanity as follows: What is the validity of any meaning and significance that is conferred on science even if it comes into relationship with all "nonscience" but does not come into relationship with religion?

Even if we leave behind the "metaphysical" concern with meaning and significance, given Robert Bellah's analysis of the modern cultural crisis, it is practically impossible for science to avoid facing the question of its relationship with religion. Bellah, in line with the insights of Weber and Durkheim, offers the following evaluation of the health of contemporary culture: "The adequacy of any ultimate perspective is its ability to transform human experience so that it yields life instead of death. Our present fragmented and disorganized culture does not rank high on that criterion" (Bellah, 1991, 245). He argues that the root cause of the "fragmented and disorganized" condition of modern culture is that (with greater fervor and zeal than the hard sciences) the soft sciences have advocated an Enlightenment myth "which views science as the bringer of light relative to which religion and other dark things will vanish away" (Bellah, 1991, 238). This modern incarnation of Manichean thought sees intuitive, emotional, imaginative, and subjective elements (epitomized in religion) as the dark forces against which war has to be waged by the forces of light represented by cognition, reason, rationality, and objectivity (epitomized in science).The disenchanted, secularized cultural condition that has emerged as a consequence of "the univocal interpretation of reality that our pseudo-scientific secular culture espouses" (Bellah, 1991, 245) has far-reaching consequences for the type of cultural being that emerges and flourishes in the disenchanted environment:

> So-called postreligious man, the cool self-confident secular man that even some theologians have recently celebrated is trapped in a literal and circumscribed reality that is classically described in religious terms as the world of death and sin, the fallen world, the world of illusion. The world of everyday reality is a socially and personally constructed world. If one confuses that world with reality itself one then becomes trapped in one's own delusions, one projects one's own wishes and fears onto others and one acts out one's own madness all the while believing that one is a clear-headed realist. Christianity, Buddhism, and other religions have long known about such delusions. They are a kind of demonic possession, for the man who believes he is most in

> control of his world is just the one most in the power of demons. (Bellah, 1991, 244f.)

This "demonic possession" often legitimates itself by appealing to the autonomy of science (or art, or politics, or economics) from all other spheres of culture and especially of autonomy from any concerns with values, ethics, and beliefs—or in simpler terms, the autonomy of the different worldly spheres from the religious sphere. This is problematic for Bellah on at least two counts:

> It is not enough to speak of the autonomy of the various spheres when the meaningfulness of our entire intellectual endeavor is in question. It is my feeling that the resources for a new integration [of religion and science] actually exist and that such a new integration would contribute much to the reunification and reinvigoration of our culture by returning to the idea of truth not only rigor but also vitality and comprehensiveness. In this way the desiccation of our culture, which is what secularization has often meant, might begin to be reversed. (Bellah, 1991, 244)

The differentiation and autonomization of the worldly spheres from the religious sphere is deeply problematic because it is the cause (more precisely the adequate cause) of the desiccation of modern culture. At the same time, this process turns it back on logical, philosophical, and scientific evidence and resources which almost necessitate that religion and science be brought into relationship. The following paragraph is an apt summation of the argument that has been made in this and the previous chapter. After summarizing the findings of Durkheim and Freud (and a little bit of Weber), Bellah notes:

> As a sociologist I am by no means prepared to abandon the work of the great consequential and symbolic reductionists. They have pointed out valid implications of religious life that were not previously understood. But I am prepared to reject their assumption that they spoke from a higher level of truth than the religious systems they studied. I would point out instead their own implicit religious positions. Most of all I am not prepared to accept the implication that the religious issue is dead and that religious symbols have nothing directly to say to us. (Bellah, 1991, 256)

The insights of Weber and Peirce taken together allow us to do four things simultaneously. Firstly, they allow us to embrace the scientific findings of not just the great scholars of religion but of all scientists great and small. Secondly, they allow us to see the limitations of that self-professed "objective" point of view that claims it can speak from a higher level of truth than the empirical reality that it is studying. Thirdly, they force us to come face to face with the religious elements that are inherent in not just scientific inquiry but in any and every cultural activity. And fourthly, they give us the ability to recognize the fact that religious symbols can still speak to us—if only we are willing to listen.

A FINAL WORD

We have seen that Weber's relational conception of science not just opens up the possibility of bringing the hard and soft sciences into relationship, it makes the establishing of this relationship almost necessary. In contrast, the disenchanting conception of science necessarily closes the possibility of establishing a relationship between the two branches of science. At the same time, we saw that both of these conceptions of science are the outcome of the exact same historical process, that is, the progress of science. We will conclude our inquiry by turning to Peirce's conception of evolution to help us understand how these two radically different possibilities can be present within one and the same process. Peirce notes: "Evolution means nothing but growth in the widest sense of that word" (1.174).. He goes on to describe "growth" in more precise terms as not "mere increase" but the simultaneous increase in diversity and regularity. From Peirce's perspective, when we look at the fourteen-billion-year history of the universe we find the continuous emergence of novel phenomena in the universe that were not present at an earlier period of time. At the same time we also witness an increase in the regularity in the universe. The further we move away from the Big Bang in terms of time (or the more that the universe evolves), the more "law-governed behavior" we notice in the universe. In short, for Peirce evolution describes the process that the universe has been undergoing for billions of years—more specifically the word "evolution" refers to that quality in the universe that describes the simultaneous growth in diversity and regularity. Faced with this particular observable quality in the universe, Peirce asks: How is evolution possible?

The two best-known answers to this question come in the form of the Darwinian and Spencerian theories of evolution. The former sees evolution being the effect of blind chance (or "sporting" in Peirce's terms), and the latter sees evolution being the effect of mechanical necessity. Peirce notes that both of these theories are inadequate for a number of reasons. He details the shortcomings of the blind-chance theory in "The Doctrine of Chances" and those of the mechanical necessity theory in "The Order of Nature." One of the most obvious shortcomings of both of these theories is almost always overlooked: "Evolution by sporting and evolution by mechanical necessity are conceptions warring against one another" (6.299). The basic reason for the incessant conflict between the two theories is that while there are facts in the universe that both of them can account for, there are other facts in the universe that they cannot account for. At a very superficial level, the observed diversity in the natural world can be seen to be the result of blind chance. But blind chance cannot account for observed regularity in the universe (except by introducing some mechanistic, "anthropological occult" doctrine like "natural selection"). Conversely, even though the observed regularity in the universe could be explained with reference to mechanical necessity, mechanical necessity cannot possibly produce the diversity and variety that characterizes the world of nature.

The shortcomings of the Darwinian and Spencerian theories of evolution can be further illustrated by showing that both of them are actually pseudo-evolutionary theories. The great triumphs of science (actually the greatest) can be collectively categorized under the heading "discovery of the laws of nature." In spite of real and irreconcilable differences on other points, the blind chance and mechanical necessity theories of evolution agree on two points regarding the "laws of nature." Firstly, "the laws of nature" are neither the product of evolution nor capable of evolution. For the mechanical-necessity theory, these laws have always existed in their present form since eternity and will continue to exist in their present form into the indefinite future. The theory displays the characteristics of the Ptolemaic/Aristotelian conception of the universe. Instead of an immutable, eternally existent universe (composed of ten spheres), we have fixed, immutable, eternally existent laws of nature. While Spencer's conception of evolution displays the logic of the Ptolemaic/Aristotelian claim that the "earth is the center of the universe," Darwin's conception of evolution embodies the logic of the creationists. Instead of the universe springing into existence, ex-nihilo, in its present form at a single instant in the past, we have (from Darwin's perspective) the laws of nature (or the laws of evolution) coming into existence, ex-nihilo, in the same form that we see them today. The fact that both Spencer and Darwin are presenting pseudo-evolutionary theories, and the implications this has for scientific inquiry, is summed up in the following observation:

> The infallibilist naturally thinks that everything always was substantially as it is now. Laws at any rate being absolute could not grow. They either were, or they sprang instantaneously into being by the sudden fiat like the drill of a company of soldiers. This makes the laws of nature absolutely blind and inexplicable. Their why and wherefore can't be asked. This absolutely blocks the road of inquiry. (1.175)

If it is indeed the case that the "laws of nature" [are] neither the product of evolution nor capable of further evolution, then "the laws of nature are absolutely blind and inexplicable" and in turn their "why and wherefore can't be asked." This is the second point that the two warring pseudo-evolutionary theories of evolution have in common—based on a priori or dogmatic principles of reasoning, they forestall scientific inquiry into the reasons, causes, and meaning of the "laws of nature." In other words both the Darwinian and

Spencerian positions make it impossible to even imagine the following question: "Why are the laws of nature as they are when they could have been otherwise?" As noted earlier, any phenomenon in the universe about which this question cannot be asked is utterly irrational and meaningless from Weber's perspective. That which is irrational and meaningless from Weber's perspective "blocks the road of inquiry" in Peirce's terms. From the foregoing discussion we can identify three of the important characteristics that can be used to distinguish a genuine (or scientific) theory of evolution from a degenerate or pseudo-scientific theory:

1. Bring diversification/spontaneity and regularity/pattern into relationship.
2. Be applicable to all phenomena in the universe, including the "laws of nature" (or "laws of history" or "laws of economics" or any other laws that are used to study phenomena in the universe).
3. Makes it possible (maybe even necessary) to raise the question "Why are the empirical phenomena in the universe and the laws that are used to study them as they are when they could be otherwise?" and offer a testable hypothesis as an answer.

Peirce offers a theory of evolution by habit-taking as an alternative to the pseudo-evolutionism of Darwin and Spencer. He sees the salient features of this alternative lying "enwrapped in the theory of Lamarck" (6.299). In the simplest terms, habit-taking makes two things possible at the same time: "It serves to establish new features [in an individual], and also to bring them into harmony with the general morphology and function of the animals and plants to which they belong" (2.300). In short habit-taking accounts, far better than the other two theories, for the simultaneous growth of novelty and diversity on the one hand and regularity and patterns on the other. For Peirce, evolution as the simultaneous growth in diversity and regularity by the tendency of habit-taking is "the general description of the action of love" (6.300). Detailing the pattern of action (or habit) that is characteristic of love, Peirce notes: "The movement of love is circular, at one and the same impulse projecting creations into independency and drawing them into harmony" (6.288). Peirce acknowledges that it is difficult to get one's mind around the logic of the claim that "the movement of love is circular." Even though this point can be illustrated using sophisticated graphing techniques, it is easier to provide an actual example from widely shared common sense: "It is fully summed up in the simple formula we call the Golden Rule" (6.288). The Golden Rule is a "universal" rule in the sense that it is found, in slightly different forms, in all of the world's major religions. At this point, Peirce offers a particular interpretation of the Golden Rule and derives a particular definition of "love": "Everybody can see that the statement of St. John is the formula of an evolutionary philosophy, which teaches that growth comes from love, from I will not say self-sacrifice, but from the ardent impulse to fulfill another's highest impulse" (6.289). Here we see that just as Peirce sees the logic of the pragmatic maxim expressed in the symbolic language of parables in Matthew 7:15–20, he finds the logic of an genuine theory of evolution expressed in the mytho-poetic language of John 3:16–21.

In light of Peirce's conception of evolution, we can go back and look at Weber's conception of progress from a new angle. Progress as disenchantment is growth in diversity and independence without corresponding growth in harmony and regularity. This leads to the fragmentation and autonomization of the different spheres of culture and the birth of inevitable and irreconcilable conflict. The split begins with the worldly spheres breaking with the religious sphere, gives birth to the conflict between religion and the worldly spheres (especially between religion and science), and then culminates in a "war of each against all" among the different worldly spheres themselves. Given the fact that evolution is the simultaneous growth of diversity and harmony, disenchantment is an example of truncated, degenerate evolution. Progress as growth in self-awareness and capacity for communication is a necessary but insufficient condition for moving toward genuine evolution. In terms of the evolution of science, we can say that Weber's work has contributed to the genuine evolution of science by pointing to the harmony and regularity between the soft and hard sciences after the two branches had become distinct from each other. In terms of the genuine evolution of culture, this would mean bringing harmony and regularity between the different spheres of culture. At a very basic level it means establishing relationships between the worldly spheres on the one hand and the religious sphere on the other. This, for its part, requires that the intellectual sphere (particularly modern science) and the religious sphere be brought into harmony. The last two chapters of this inquiry have put forth the argument that the insights of Weber and Peirce, taken separately, contain a wealth of theoretical capital to move in this direction. When the work of these two past masters is brought into relationship, the value of the theoretical capital becomes even more apparent.

Summarily stated the value of the theoretical capital that is contained in the work of Weber and Peirce is that it turns two different but intimately related divides into relationships; the divide between the hard and soft science and the divide between science and religion. The implications of this transformation from division (and the resulting conflict) into relationship (and resulting harmony) are far reaching for both modern science and modern culture. On this point we will let Percy and Bellah have the last word. Percy notes that there are a number of different consequences for Peirce's "discovery that precisely that which is distinctive in human behavior, language, art, thought itself, is not accounted for by the standard scientific paradigm which has been sovereign for three hundred years" (Percy, 1991, 288). That which "the standard scientific paradigm" cannot account for is a creature (an interpretant, interpreter, ego, I) that "throws together word and thing"—a creature who uses symbols. The first implication is that Peirce's discovery points to "the shape of a new science" (Percy, 1991, 288) or as we have seen a more scientific conception of science. This more scientific conception of science for its part reveals that the "words, symbols, and the things symbolized are subject to *norms*, something new in the world. They can be fresh and grow stale. Words can tell the truth or lie" (Percy, 1991, 289). The discovery that human beings use symbols to describe a growing and evolving world around them opens up new horizons of inquiry—especially in the form of the ability to asks questions that cannot be asked otherwise.

This opening leads Percy to make the following prediction: "All humanists, even novelists, are entitled to make prophecies. Here is the prophecy: The [hard] scientist of the future will be able to make sense of the following sort of sentence which presently makes no sense to him whatsoever: *There is a difference between the being-in-the-world of the scientist and the being-in-the-world of the layman*" (Percy, 1991, 290). With the anthropology that is implicit in Peirce's semiotics and hermeneutics in hand "one might even explore its openness to such traditional Judeo-Christian notions as man falling prey to the worldliness of the world, and man as pilgrim seeking his salvation" (Percy, 1991, 291). In sum, Percy sees a future where Peirce's discovery of the role and use of the symbol in the human interaction with the world opens up the possibility of shared language which makes it possible for scientists and theologians, believers and non-believers to explore aspects of empirical reality and ask investigative questions that could not be explored or investigated before.

Taken together the insights of Weber and Peirce give us a coherent and comprehensive theory of what Bellah calls "symbolic realism":

> The canons of empirical science [i.e., the hard sciences] apply primarily to symbols that attempt to express the nature of objects, but there are nonobjective symbols that express the feelings, values, and hopes of the subjects, or that organize and regulate the flow of interactions between subjects and objects, or that attempt to sum up the whole subject-object complex or even point to the context or ground of that whole. These symbols, too, express reality and are not reducible to empirical propositions. This is the position of symbolic realism. (Bellah, 1991, 252)

Weber's notion of ideal-type and Peirce's notion of icon, index, symbol provide the conceptual tools to capture "feelings, values, and hopes of the subjects and organize and regulate the flow of interactions between subjects and objects." And Weber's conception of the relationship between understanding, explanation, and interpretation and Peirce's notion of studying phenomena at class-8, -9, and -10 signs "sum up the whole subject-object complex." Speaking as a practicing social scientist who is fully committed to "the canons of scientific objectivity [and] value neutrality" (Bellah, 1991, 256), Bellah notes that the theory of symbolic realism "is the only adequate basis for the social scientific study of religion" (Bellah, 1991, 253). If both the theologian and the scientist approach their subject matter through the lens of symbolic realism "then we are in a situation where for the first time in centuries theologian and secular intellectual can speak the same language. Their tasks are different but their conceptual framework is shared. What this can mean for the reintegration of our fragmented culture is almost beyond calculation" (Bellah, 1991, 253).

In conclusion, the insights of Weber and Peirce allow us to see clearly that the "iron cage" that is the "fate of our times" is neither the result of any "law of nature" nor anything inherent in "human nature." It is the manifestation of arrested development. In other words, the death of meaning, the perpetual and irreconcilable conflict between the different spheres of culture, and the reduction of human cultural activity to "a mere sporting contest" that is characteristic of life in the "iron cage" is the result of a truncated, degenerate form of evolution. Evolution in its full, genuine sense contains other cultural possibilities for the future. One of those other possibilities is the rediscovery of meaning (*Sinn*) in the universe, establishing relationships and harmony between the different spheres of culture (especially the intellectual and religious spheres), and engaging in purposeful cultural activity where even "sporting events" have serious ends. Assuming that we are living in a genuinely evolving universe, a whole host of different factors will have their say in terms of the future that the evolutionary process leads to. The Weberian conception of science and the Peircean conception of philosophy show that the beliefs and values of the cultural beings who are part of the evolving historical constellation will play an especially important role in determining which of the future cultural possibilities are actualized rather than others.

NOTES

1. This becomes a point of significance in Western intellectual history in light of Oswald Spengler's analysis of the relationship of modern, Western culture and classical Greek culture. Spengler argues:

> In all history, so far, there is no second example of one Culture paying to another Culture long extinguished such reverence and submissions in matters of science as ours has paid to the Classical. It was very long before we found courage to think our proper thought. But though the wish to emulate the Classical was constantly present, every step of the attempt took us in reality further away from the imagined ideal. The history of Western knowledge is thus one of *progressive emancipation* from Classical thought. (Spengler, 1959, 78)

If the spirit of modern Western culture is indeed thoroughly anticlassical, as Spengler argues, then Weber's reflections on the methodology of scientific inquiry become a watershed event in Western intellectual history. Weber's reflections are one of the first and most radical breaks with classical metaphysics in the soft sciences. Spengler argues that the break of modern European culture from the classical Greek heritage is the result of purely European genius and indigenous European developments. Muhammad Iqbal (d. 1938) agrees with Spengler's claim that the spirit of modern Western culture is thoroughly anticlassical (Iqbal, 2012, 114). Summarily stated, the classical conception of the universe is that of a closed, completed, static universe in which all movement is nothing more than movement toward a predetermined end. In such a universe it is possible to attain certain, exact, and universal knowledge. The modern conception of the universe is that of a dynamic, open, evolving universe in which movement toward genuine novelty and new horizons is possible.

While both Spengler and Iqbal agree that the birth and evolution of modern culture marks a break with classical Greek culture, Iqbal challenges Spengler's claim that non-European elements played no causal role in the birth of the modern spirit. Iqbal locates the birth of the spirit of modern culture in the event of the revelation of the Qur'an. Iqbal argues that "the spirit of the Qur'an was essentially anti-classical" (Iqbal, 2012, 3) in terms of the view of the universe, the type of knowledge that can be attained about the universe, and the place of the human being in the universe. With reference to the findings of the historians of science in the early part of the 20th century, Iqbal puts forth the argument that the ideas of Muslim thinkers inspired by the Qur'an played a causal role in the birth of the spirit of modern culture (Iqbal, 2012, 99–115).

Leaving aside the debate between Spengler and Iqbal on the origins of the spirit of modern culture, the one point on which they agree presents us with a most intriguing line of inquiry. If the spirit of modern culture is indeed anticlassical as both of them claim, and if the spirit of the Qur'an is anticlassical as Iqbal claims, then there should be significant points of overlap, affinity, and relationship between Weberian sociology on the one hand and the teachings of the Qur'an (as interpreted by Iqbal) on the other. The fact that Iqbal earned the sobriquet Rumi II during his lifetime evidences that he is not a marginal figure in Muslim intellectual history. At the same time, Weber stands firmly within the tradition of Western thought. Hence any overlap between Iqbal's reading of the Qur'an and Weberian sociology would contribute to conversation, interaction, and exchange between Islam and the West in the twenty-first century from perspectives that are located squarely within the two traditions that are in conversation.

References

Adler, Alfred. 1991. "Mathematics and Creativity." In *The World Treasury of Physics, Astronomy and Mathematics*, edited by Timothy Ferris. New York: Little, Brown.

Agevall, Ola. 2009. "A Science of Unique Events: Max Weber's Methodology of the Cultural Sciences." In *The Protestant Ethic and the Spirit of Capitalism*, edited by Richard Swedberg, translated by Talcott Parsons. New York: W.W. Norton. Pp. 155–179.

Akhtar, Junaid, Basit Bilal Koshul, and Mian Muhammad Awais. 2013. "A Framework for Evolutionary Algorithms Based on Charles Peirce's Evolutionary Semiotics." *Information Sciences* 236: 93–108.

Albrow, Martin. 1987. "The Application of the Weberian Concept of Rationalization to Contemporary Conditions." In *Max Weber, Rationality and Modernity*, edited by Scott Lash and Sam Whimster. London: Allen & Unwin. Pp. 164–182.

Alexander, Jeffery. 1982. *The Classical Attempts at Theoretical Synthesis: Max Weber*. Berkeley: University of California Press.

———. 1987. "The Dialectic of Individuation and Domination: Weber's Rationalization Theory and Beyond." In *Max Weber, Rationality and Modernity*, edited by Scott Lash and Sam Whimster. London: Allen & Unwin. Pp. 185–206.

Bellah, Robert. 1991. *Beyond Belief: Essays on Religion in a Post-Traditional World*. Berkeley: University of California Press.

Bendix, Reinhart. 1962. *Max Weber: An Intellectual Portrait*. New York: Anchor.

Berger, Peter.1992. "Sociology: A Disinvitation?" *Society* 36 (1): 12–18.

Bourdieu, Pierre. 1987. "Legitimation and Structured Interests in Weber's Sociology of Religion." In *Max Weber, Rationality and Modernity*, edited by Scott Lash and Sam Whimster. London: Allen & Unwin. Pp. 119–136.

Brubaker, Rogers. 1984. *The Limits of Rationality: An Essay on the Social and Moral Thought of Max Weber*. London: George Allen & Unwin.

Ciaffa, Jay. 1998. *Max Weber and the Problem of Value-Free Social Science: A Critical Examination of Wertulteilsstreit*. Cranbury, NJ: Associated University Press.

Deely, John. 1994. *New Beginnings: Early Modern Philosophy and Postmodern Thought*. Toronto: University of Toronto Press.

Dewey, John. 1920. *Reconstruction in Philosophy*. New York: Henry Holt.

Diggins, John. 1996. *Max Weber: Politics and the Spirit of Tragedy*. New York: Basic.

Feyerabend, Paul. 1984. *Against Method: Outlines of an Anarchistic Theory of Knowledge*. London: Verso.

Gunnell, John. 2007. "The Paradox of Social Science: Weber, Winch and Wittgenstein." In *Max Weber's 'Objectivity' Reconsidered*, edited by Laurence McFalls. Toronto: University of Toronto Press. Pp. 58–88.

Heidelberger, Michael. 2010. "From Mill via von Kries to Max Weber: Causality, Explanation, and Understanding." In *Historical Perspectives on Erklären and Verstehen*, edited by U. Feest. New York: Springer. Pp. 241–265.

Hekman, Susan. 1979. "Weber's Concept of Causality and the Modern Critique." *Sociological Inquiry* 49 (4): 67–76.

———. 1983. "Weber's Ideal-type: A Contemporary Reassessment." *Polity* 16 (1): 119–137.

Hindess, Barry. 1987. "Rationality and the Characterization of Modern Society." In *Max Weber, Rationality and Modernity*, edited by Scott Lash and Sam Whimster. London: Allen & Unwin. Pp.137–153.

Holton, Gerald. 1973. *Thematic Origins of Scientific Thought: From Kepler to Einstein*. Cambridge, MA: Harvard University Press.

Huff, Toby. 1984. *Max Weber and the Methodology of the Social Sciences*. New Brunswick, NJ: Transaction.

Iqbal, Muhammad. 2012. *The Reconstruction of Religious Thought in Islam*. Stanford, CA: Stanford University Press.

Jenson, Robert. 1997. *Systematic Theology. Vol. 1, The Triune God*. Oxford, UK: Oxford University Press.

Kadushin, Max. 1952. *The Rabbinic Mind*. New York: Bloch.

Kalberg, Stephen. 1980. "Max Weber's Types of Rationality: Cornerstones for the Analysis of Rationalization Process in History." *American Journal of Sociology* 85 (5): 1145–1179.

———. 1994a. *Max Weber's Comparative-Historical Sociology*. Chicago, IL: University of Chicago Press.

———. 1994b. "Max Weber's Analysis of the Rise of Monotheism: A Reconstruction." *British Journal of Sociology* 45 (4): 563–583.

———. 1996. "On the Neglect of Weber's Protestant Ethic as a Theoretical Treatise: Demarcating the Parameters of Postwar American Sociological Theory." *Sociological Theory* 14 (1): 49–70.

Koshul, Basit. 2005. *The Postmodern Significance of Max Weber's Legacy*. New York: Palgrave Macmillan.

Krampen, Martin. 1981. "Phytosemiotics." *Semiotica* 36 (3/4): 187–209.

Kull, Kalevi. 2000. "An Introduction to Phytosemiotics: Semiotic Botany and Vegetative Sign Systems." *Sign Systems Studies* 28: 326–350.

McFalls, Laurence, et al. 2007. "Conclusion: The 'Objectivist' Ethic and the 'Spirit' of Science." In *Max Weber's "Objectivity" Reconsidered*, edited by Laurence McFalls. Toronto: University of Toronto Press. Pp. 351–373.

Mertz, Elizabeth. 2007. "Semiotic Anthropology." *Annual Review of Anthropology* 36 (1): 337–353.

Mitzman, Arthur. 1970. *The Iron Cage: An Historical Interpretation of Max Weber*. New York: Knopf.

Nietzsche, Friedrich. 1974. *The Gay Science: With a Prelude in Rhymes and an Appendix in Songs*. Translated by Walter Kaufmann. New York: Vintage.

———. 1998. *Philosophy in the Tragic Age of the Greeks*. Washington, DC: Regnery Publishing Inc.

Oakes, Guy. 1975. "Introductory Essay." In *Roscher and Knies: The Logical Problems of Historical Economics*, by Max Weber, translated by Guy Oakes. New York: Free Press. Pp. 1–49.

———. 1988. *Weber and Rickert: Concept Formation in the Cultural Sciences*. Cambridge, MA: MIT University Press.

Ochs, Peter. 1998. *Peirce, Pragmatism and the Logic of Scripture*. Cambridge, UK: Cambridge University Press.

Peirce, Charles. 1931–1958. *Collected Papers of Charles Sanders Peirce*. 8 vols. Vols. 1–6 edited by Charles Hartshorne and Paul Weiss. Vols. 7–8 edited by Arthur W. Burks. Cambridge, MA: Harvard University Press.

Penrose, Roger. 1994. *Shadows of the Mind: A Search for the Missing Science of Consciousness*. New York: Oxford University Press.

Percy, Walker. 1991. *Signposts in a Strange Land*. New York: Farrar, Straus & Giroux.

Raposa, Michael. 1989. *Peirce's Philosophy of Religion*. Bloomington, IN: Indiana University Press.
Ringer, Fritz. 2000. *Max Weber's Methodology: The Unification of the Cultural and Social Sciences*. Cambridge, MA: Harvard University Press.
Schluchter, Wolfgang. 1979. "The Paradox of Rationalization: On the Relation of Ethics and World" and "Value Neutrality and the Ethic of Responsibility." In *Max Weber's Vision of History: Ethics and Methods*, edited by Guenther Roth and Wolfgang Schluchter. Berkeley: University of California Press. Pp. 11–116.
Sebeok, Thomas. 1994. *Signs: An Introduction to Semiotics*. Toronto: University of Toronto Press.
Sheriff, John. 1989. *The Fate of Meaning: Charles Peirce, Structuralism and Literature*. Princeton, NJ: Princeton University Press.
———. 1994. *Charles Peirce's Guess at the Riddle: Grounds for Human Significance*. Bloomington, IN: Indiana University Press.
Sica, Alan. 2004. *Max Weber and the New Century*. New Brunswick, NJ: Transaction.
Singer, Milton. 1984. *Man's Glassy Essence: Explorations in Semiotic Anthropology*. Bloomington, IN: Indiana University Press.
Spengler, Oswald. 1959. *The Decline of the West*. London: George Allen & Unwin.
von Uexküll, Jakob. 2010. *A Foray into the World of Humans and Animals: With a Theory of Meaning*. Minneapolis, MN: University of Minnesota Press.
Weber, Max. 1946a. "Religious Rejections of the World and Their Directions." In *From Max Weber,* edited by H. H. Gerth and C. Wright Mills. New York: Oxford University Press. Pp. 329–359.
———. 1946b. "Science as a Vocation." In *From Max Weber*, edited by H. H. Gerth and C. Wright Mills. New York: Oxford University Press. Pp. 129–156.
———. 1946c. "The Social Psychology of World Religions." In *From Max Weber*, edited by H. H. Gerth and C. Wright Mills. New York: Oxford University Press. Pp. 267–322.
———.1949. *Max Weber on the Methodology of the Social Sciences*. Edited by Edward A. Shils and Henry A. Finch. Glencoe, IL: Free Press.
———. 1968. *Economy and Society*. Edited by Guenther Roth and Claus Wittich. New York: Bedminster.
———. 1975. *Roscher and Knies: The Logical Problems of Historical Economics*. Translated by Guy Oakes. New York: Free Press.
———. 1977. *Critique of Stammler*. Translated by Guy Oakes. New York: Free Press.
———. 1981. "Some Categories of Interpretive Sociology." *Sociological Quarterly* 22: 151–180.
———. 2002. *The Protestant Ethic and the Spirit of Capitalism*. Translated by Stephen Kalberg. Los Angeles: Roxbury Press.
Weiss, Johannes. 1987. "On the Irreversibility of Western Rationalization and Max Weber's Alleged Fatalism." Translated by Bruce Allen and Sam Whimster. *Max Weber, Rationality and Modernity*. Edited by Scott Lash and Sam Whimster. London: Allen & Unwin. Pp. 154–163.
Westphal, Merold. 1993. *Suspicion and Faith: The Religious Uses of Modern Atheism*. Grand Rapids, MI: Wm. B. Eerdmans.
Whimster, Sam, and Scott Lash, eds. 2006. *Max Weber, Rationality and Modernity*. London: Routledge.
Wittenberg, Erich. 1989. "The Crisis of Science in Germany in 1919." In *Max Weber's "Science as a Vocation,"* edited by Peter Lassman and Irving Velody, with Herminio Martins. London: Unwin Hyman. Pp. 122–137.
Witzany, Günther. 2005. "From Biosphere to Semiosphere to Social Lifeworlds: Biology as an Understanding Social Science." *tripleC: Cognition, Communication, Co-operation* 3 (2): 51–74.
———. 2008. "The Biosemiotics of Plant Communication." *American Journal of Semiotics* 24 (1): 39–56.

Index

About the Author

Basit Bilal Koshul received his first PhD in 2003 from Drew University, in the area of religion and society specializing in the sociology of religion. After teaching at Concordia College in Moorhead, MN for four year he joined the Department of Humanities and Social Sciences at the Lahore University of Management Sciences in Lahore, Pakistan, in 2006. He is currently an associate professor there teaching sociology and philosophy. He completed a second PhD in 2011 from the University of Virginia in religious studies specializing in theology, ethics, and culture. His areas of interest include the relationship between religion and modernity, philosophy of science, philosophy of religion, the sociology of culture, and the contemporary Islam–West encounter. He is especially interested in using the post-critical methodology of scriptural reasoning to study and integrate the ideas of Muhammad Iqbal, Charles Peirce, and Max Weber.

www.ingramcontent.com/pod-product-compliance
Lightning Source LLC
LaVergne TN
LVHW101446200926
848483LV00001B/69